Design Characteristics of a Space Elevator Earth Port

Prepared by the

International Space Elevator Consortium

Fall 2015

Authors: Vern Hall, Peter Glaskowsky, Sandee Schaeffer

Editor: Skip Penny

Design Characteristics of a
Space Elevator Earth Port

Front Cover Illustration:

chasedesignstudios.com

Published by Lulu.com

Pete.swan@isec.org

ISBN 978-1-329-91060-7

The International Space Elevator Consortium

16991 McGill Road, Saratoga, CA 95070 / http://www.isec.org

Printed in the United States of America

Preface

The Earth terminus of the space elevator has been discussed in general terms since the beginning of the concept. In fact, Tsiolkovsky started with a very tall tower on the Earth and went upward with his ideas. Co-inventors Yuri Artsutanov (1960) and Jerome Pearson (1974) developed the modern space elevator and went down to Earth from geosynchronous, with a quick discussion of the Earth terminal. Since then, there have been many discussions and assertions. As such, a definitive look at the lower terminus of the space elevator seems timely.

This led us (International Space Elevator Consortium –ISEC) to the selection of the Marine Node description as the 2015 yearly topic. This topic was selected to concentrate efforts toward a meaningful understanding of the Marine Node, in this case as a study and subsequent report. This ISEC focus enables the community to contribute towards a designated goal of describing the requirements for building and operating the marine node. In addition, this allows participation in various activities such as ISEC journal articles, student contest topics, conference theme, and a major study. This topic was initiated in September 2014 with publication planned for the spring of 2016.

During this study, the name of the supporting activities at the lower end of the space elevator has transitioned from the accepted "Marine Node" to the newly agreed upon "Earth Port." This new term is more definitive and describes any facility near the equator and the multiple functions inherent in operating a transportation hub.

This report represents the culmination of efforts by multiple contributors describing the functions and operations of the Earth Port and how it interacts with other space elevator nodes. The challenge was to bring in a fresh perspective on an Earth Port by leveraging experts in many fields beyond those normally associated with space elevators. One example of this professional expertise was the application of 45 years of knowledge about a specific transportation infrastructure, namely the Port of Los Angeles. As a result of the contributions from a variety of professional backgrounds, this study should reflect a more realistic perspective of the activities at the Earth Port of a future space elevator.

This study is the latest in a series of ISEC year-long efforts, including

> (1) 2010, <u>Space Elevator Survivability, Space Debris Mitigation</u>,
>
> (2) 2012, <u>Space Elevator Concept of Operations</u>,
>
> (3) 2013, <u>Design Considerations for Space Elevator Tether Climbers</u>, and
>
> (4) 2014, <u>Space Elevator Architectures and Roadmaps</u>.

The authors of this report wish to thank: (1) the members of ISEC for their support, (2) contributors to this report for their dedicated efforts, and (3) the attendees of the 2015 International Space Elevator Conference.

Signed: *Robert E "Skip" Penny, Jr*

Vice President ISEC

Executive Summary

This study report provides the International Space Elevator Consortium's (ISEC) view of the Earth Port (formerly known as the Marine Node) of a Space Elevator system. The Earth Port:

- serves as a mechanical and dynamical termination of the space elevator tether, providing reel-in/reel-out capability and position management in order to deal with tension, wind, current and debris avoidance; it may also serve as a satellite terminus platform;
- serves as a port for receiving and sending Ocean-going Vessels (OGVs); the OGVs that come and go from the Earth Port will move tether climbers, payloads, supplies and personnel;
- provides landing pads for helicopters from the OGVs;
- serves as a facility for attaching and detaching payloads to and from tether climbers and attaching and detaching climbers to and from the tether;
- provides tether climber power for the 40 km above the Floating Operations Platform (FOP);
- provides food and accommodation for crew members as well as power, desalinization, waste management and other such support.

Early in the report, the rationale for a space elevator is described. In addition, the accepted assessment of the first Earth Port location is shown to be near the equator in the Pacific. Other locations are possible. One attractive location is Howland Island, described in Appendix F. The bottom line is that the space elevator will create a transportation infrastructure that will provide revolutionary access to space routinely, inexpensively, safely, daily, and with large payloads.

Next, the Functional Requirements are presented to show a comprehensive enumeration of the various features and functions of the Earth Port. This section sets the stage for a broad spectrum of designs for future space elevators.

The main focus of the report is Section 4 which offers an in-depth description of the Earth Port deriving from the requirements discussed previously. There were many aspects within this section that intrigued the team. One continuing question was how comprehensive would the Floating Operations Platform need to become?

Section 5 poses a possible organizational structure including staffing estimates. The allocation of tasks among the people assigned to the Earth Port continually matured during the year-long study.

Section 6 describes a notional "Day in the Life" of a crew member. It is hoped that this section will provide the designers with an insight into the daily work at the Earth Port and to consider the necessary human factors analyses.

Section 7 is a recap of the Earth Port development Roadmap which developed during the 2014 year-long study.

Section 8 sets forth our conclusions and recommendations.

The contributors concluded:

1. **Design and construction of the Earth Port will be a straightforward extension of today's practices**. As the Earth Port consists of many components, there will be a systems engineering design aspect that will ensure consistency between the Floating Operations Platform (FOP) and logistics flow of support and mainline mission equipment. The Earth Port has a tremendous amount of historical experience to leverage. The design of floating platforms goes back thousands of years and the ability to operate from them is mature. In addition, there are many parallel developments of operations centers that can help design the FOP and any satellite terminus. The roadmap, accomplished for the Earth Port in the 2014 ISEC study (Space Elevator Architecture and Roadmaps), shows that development can be accomplished in a linear manner with current technologies.

2. **Operations at the Earth Port will leverage over a hundred years of combined experience**. This expertise and knowledge will come from off-shore drilling operations and naval operations around the world. The FOP will likely be a modified drilling platform to support the combined needs of securing space elevator tethers and enabling the processing of payloads through its transportation hub. Each of the major components of the Earth Port will require support from experienced personnel who have worked inside off-shore drilling platforms and transportation hubs.

3. **Operation and maintenance costs appear to be reasonable**. By using parallel constructs, the estimate of staffing and platform sizes has enabled this team to estimate the costs. The key here is to ensure that the operations are routine and established early in the development of our space elevator concept. Most of the processes must be straightforward to ensure that the routine movement of payloads and tether climbers is enabled in a rapid manner, supporting the needs of customers. There will be a business center at the Headquarters that is responsible for ensuring the costs are estimated correctly and then adhered to.

Recommendation: As each of the International Space Elevator Consortium's year-long studies finishes, it is obvious that we have just initiated the discussions on the topic. As such, this team recommends that efforts continue to refine the ideas in this report to help reduce risk for future development.

Table of Contents

List of Figures

1.   Introduction

The modern day space elevator concept was defined by Dr. Brad Edwards during the first few years of the 21st century. Most recently, the International Academy of Astronautics (IAA) published a report[1] updating the body of knowledge. In addition, the Obayashi Corporation has published papers and PowerPoint presentations laying out their design characteristics of a space elevator system.

With over 40 years of experience in space, the industry seems eager to support the idea that an inexpensive transportation infrastructure to space could be constructed as soon as the materials are available. The resulting operations would be a mix of historic maritime transportation approaches and unique day-to-day space operations.

This report addresses the Earth Port of a space elevator system with emphasis on the design characteristics that would lead to a near-term space elevator. Figure 1-1 is a schematic of the system we envision showing its major components and the location of the Earth Port relative to them.

---

[1] Space Elevators: An Assessment of the Technological Feasibility and the Way Forward, P. Swan, D. Raitt, C. Swan, R. Penny and J. Knapman, International Academy of Astronautics, October 2013

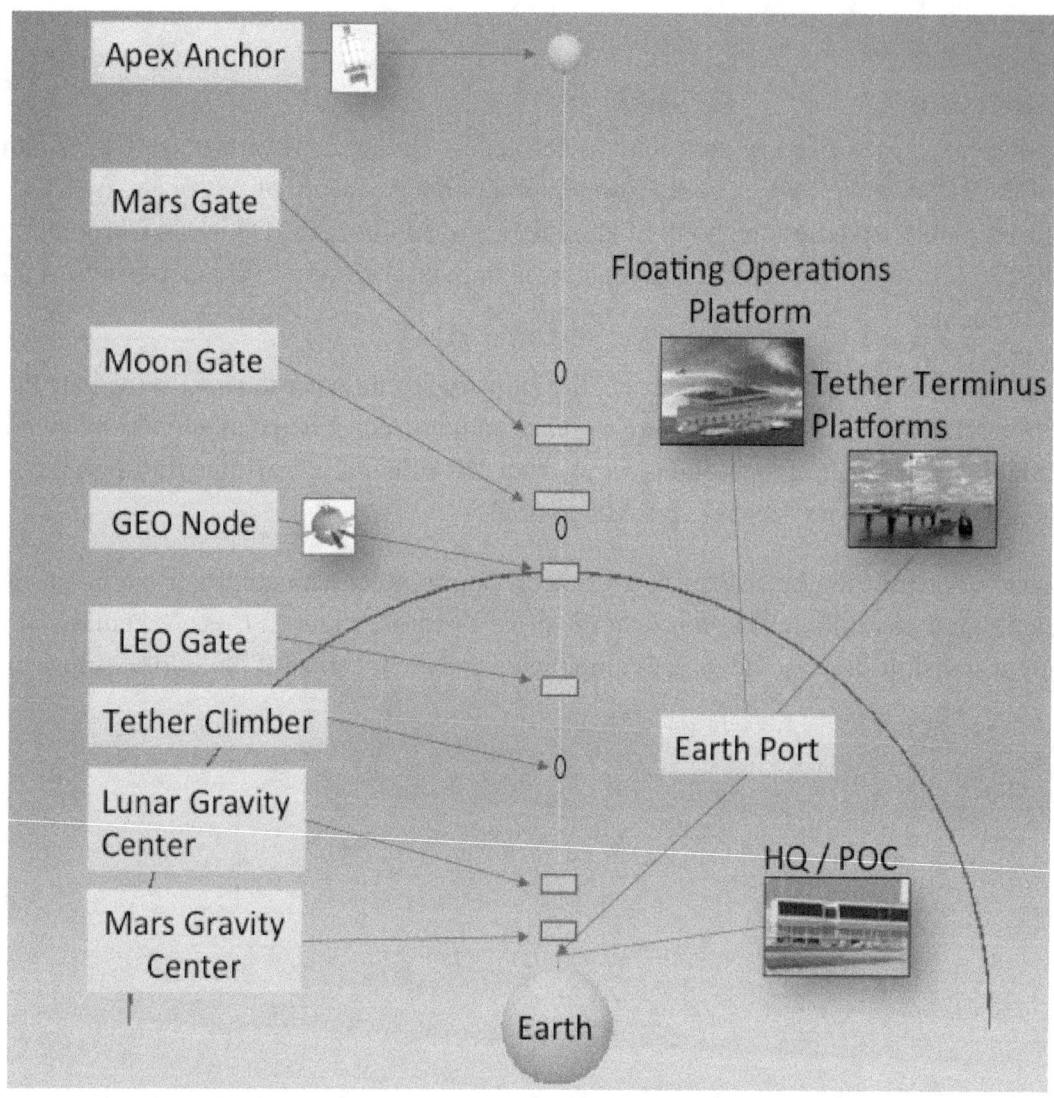

Figure 1-1 Space Elevator System [image by Nick Regnier]

LEO, Moon and Mars Gates represent locations along the space elevator where payloads could be released to those destinations. The Apex Anchor, GEO Node, Climber and Earth Port are critcal elements in the physical design of the space elevator. The Lunar and Mars Gravity Centers are proposed research stations exploiting the effective gravity at those locations.

2. Why a Space Elevator and Where Should the Earth Port Be Located

2.1 Why a Space Elevator?

Ted Semon, past president of ISEC (2009-2013), explained why space elevators should become an operational space transportation infrastructure. His review of the advantages of a hypothetical space elevator over conventional rockets is paraphrased here.

• The space elevator is scalable. If and when space elevators become possible, a sufficiently large number of them could be built so that hundreds of tons of payload could be sent from earth to space every day.
• It is a transportation infrastructure, like the trans-continental railroad.
• Similar to riding on a high-speed train, the space elevator would remove worry about cushioning cargo against high-g forces and the "shake, rattle and roll" that accompany rocket launches.
• Pollution generated by the space elevator will likely be very small, far less than that of rockets.
• The space elevator is expected to have a very small failure rate, much less than the 1-3% failure rate of modern rockets.
• Specific advantages of the space elevator include:
    • low cost access to Geostationary Earth orbit (GEO) and beyond, estimated to be in the neighborhood of $100 per kg,
    • rapid expansion of commercial industry in space made possible by cheap access to orbit,
    • expansion of the Earth-based businesses that would provide the necessary new technologies,
    • the possibility of solar power satellites at GEO to deliver cheap energy to Earth,
    • the enabling of space mineral resource mining on the Moon, Mars and near-Earth asteroids, and in order to aid mining,
    • lunar space elevators.[2]

2.2 Where should the Earth Port be located?

This study focuses on the Earth Port as a hub for the full transportation infrastructure. The assumption going into this study was that the Earth Port would be located on or near the equator in the open Pacific Ocean west of the Galapagos as shown in Figure 2-1. Many other locations near the equator will work also. The rationale for the selection is twofold. The first is the legal freedom given to ships, stations, and humans in the open ocean, as an international zone with the Law of the Sea dominating. These strengths leverage the heritage of the sea with its own laws and history of political insulation. Secondly, the location of facilities in the equatorial Pacific will enhance the safety and operational aspects of the Earth Port because of the mild weather conditions (wind, waves, and lightning) in the area, and the reduced likelihood of collateral damage in case of accidents. Each of these factors leads to the conclusion that the open ocean of the equatorial Pacific is a good choice for the location of the Earth Port.

---

[2] Ibid.

In addition, sea-based infrastructure prospers around the world with logistical strengths for long distance transportation, simplicity, and proven technologies. There are tremendous areas in the mid-Pacific that are open and seldom have human contact. Most ships travel in known routes that are very far from the suggested location. Various factors that contribute to the necessary features and functions of the Earth Port are addressed in Section 3.

For these reasons the IAA study[3] recommended that the Earth Port be located at latitude 5° S (approximately) and longitude 100° W, as shown below.

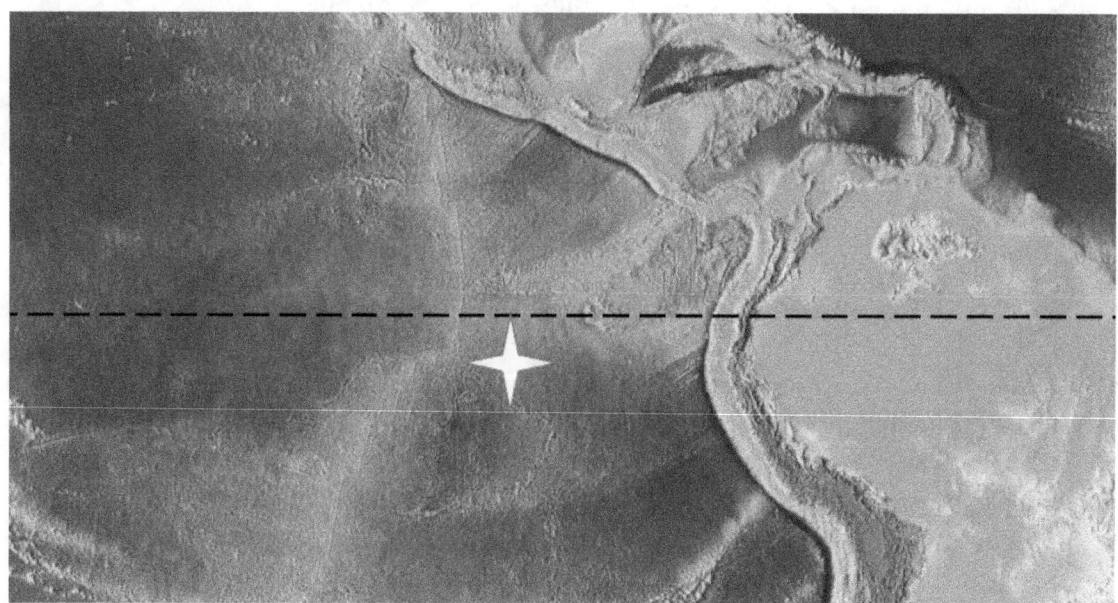

Figure 2-1 Recommended location in Pacific

³ Ibid.

3. Earth Port Functional Requirements

In the ISEC baseline space elevator system, the tether terminates within the Earth Port, a sea-based structure that performs multiple functions. We will describe the functional requirements for the Earth Port in terms of the services it provides to the various elements of the complete system.

In practice, we anticipate a single Earth Port will consist of Floating Operations Platforms (FOPs) and multiple, separate terminals, each with an attached tether. For best efficiency, each tether will be devoted to upward or downward transportation. There may be additional floating platforms without tethers to provide operational support. For convenience, within this Section, we will refer to the Earth Port as if it were a single entity even though it may consist of several structures.

The key elements of the overall space elevator system are listed below and then discussed individually in a functional walk-through.

- Tether, including reel, service facility, climbers and shroud
- Cargo processing
- Transportation, including Ocean-going Vessels (OGVs) and air vehicles
- Logistics
- Storage
- FOP positioning
- Utilities: power, water
- Maintenance facilities
- Safety
- Security
- Communications
- Offices
- Quarters

3.1 Tether

The tether itself consists of a ribbon of Carbon Nanotube (CNT) material. It extends from the Earth Port up through GEO Node and on to the Apex Anchor, which is located a nominal 100,000 km above the surface of the Earth.

3.2 Tether reel

The tether is not rigidly fixed to the Earth Port; instead, it terminates in a reel that holds extra tether material. The reel allows the tension on the tether to be adjusted. It will also assist in tether positioning and dynamic stabilization. Lateral tether positioning at the Earth Port, which will likely be required to help mitigate threats of damage from orbiting debris, will be addressed below under Positioning. The power rating for the tether reel subsystem will depend on tether tension as well as the requirements for acceleration and peak speed.

3.3 Tether service facility

The tether will also need to be cleaned, inspected, and maintained. A portion of the tether can be serviced within the Earth Port by reeling the tether in and out. The portion of the tether which can be serviced within the Earth Port depends upon the capacity of the reel. The rest of the tether must be serviced by tether climbers. There may be tether service subsystems on all tether climbers, and there may be special-purpose tether climbers just to perform tether servicing.

3.4 Tether climbers

Most tether climbers will be used primarily to move payloads along the tether. ISEC anticipates that most traffic along the tether will be in the upward direction. It is likely that there will be different types of tether climbers, each specializing in moving a particular kind of payload, or in moving payload along a different portion of the tether. Repair and maintenance climbers will also be necessary.

For example, the requirements for tether climbers operating within the atmosphere will likely be significantly different from climbers operating only in the vicinity of the Geosynchronous Node at 35,789 km altitude, where the effects of the atmosphere are negligible. Gravitational forces at that altitude are greatly reduced, so climbers could use relatively small motors. [See ISEC Position Paper 2013-1 "Design Considerations for Space Elevator Tether Climbers" for more information about tether climber characteristics.]

3.5 Tether climber shroud

Another key difference in design characteristics is that tether climbers are expected to require a protected environment while traveling within the atmosphere. To meet this requirement, ISEC anticipates that tether climbers will be contained within a protective shroud between the Earth Port and 40 km altitude.

These shrouds may have their own drive motors (which would relieve the tether climber of the need to lift the mass of the shroud and allow the shroud to travel upward on its own to meet a descending climber), and will be powered through an electrical cable. The cable will have its own reel within the Earth Port. We estimate that the 40 km power cord will weigh about 2.4 metric tons.

The shroud drive motor subsystem will be capable of lifting the shroud, the tether climber, and power cable. It could be that the shroud has no motor and is lifted by the climber. The lift capacity will be determined by design engineering and the allocation of lift capacity will be the subject of system engineering trade studies.

The process of preparing and dispatching climbers can be broken down into several phases:

- receiving cargo, including tether climbers and payloads, from OGVs,

- storing cargo until it is needed,
- preparing climbers and payloads,
- loading payloads into climbers,
- integrating the climber with the shroud,
- attaching the integrated vehicle to the tether,
- testing and dispatching the vehicle,
- supporting the vehicle until it is handed off to the next node and
- cargo processing.

The process for returning climber/shrouds would be essentially the inverse of the above list.

We use "cargo" as a generic term to refer to all materials, including climbers, payloads, consumable items, and so on, while they are in the process of being moved or stored as opposed to being used for their intended purpose. Cargo on returning climber/shrouds would be any reusable equipment as well as recovered satellites and space debris.

3.6 Transportation

OGVs operating as part of the space elevator system may be designed specifically for this application, including optimizations to facilitate interactions with the Earth Port. Such design would incorporate many of the existing, established technologies used in current ship transports.

Most cargo will arrive or depart on OGVs. These vessels must be provided with a protected moorage while they are being loaded and unloaded.

OGVs may also be used to transport personnel and may include high-speed ferries like those used for other passenger transport applications.

High-priority cargo and some personnel traveling to and from the Earth Port will be transported by air vehicles. Aircraft may also be used for emergency transport when needed.

We anticipate that the Earth Port will support seaplanes and Vertical Take Off and Landing (VTOL) vehicles such as helicopters. Tie-downs, hangar space, moorage, refueling and maintenance facilities will be provided for air vehicles as required.

3.7 Logistics

Cargo movement within the Earth Port will require material-handling equipment such as cranes, forklifts, conveyors, and so on. The Earth Port must be prepared to handle unusually large and delicate items. Some of these items are highly predictable (tether climbers, climber shrouds) and some are unpredictable (payloads).

Also note that space elevator payloads may include hazardous materials such as rocket propellants. We anticipate that some payloads will depart the tether (typically at altitudes from 23,412 km to 46,722 km) and use rockets to deliver them to their final destinations. Those upper stages must be lifted by the space elevator, so they will transit the Earth Port just like any other payload.

3.8 Storage

The Earth Port will include a substantial amount of storage space. Multiple complete tether climbers will be stored in various stages of readiness. One or more tether climber shrouds will be stored, and each of these must be larger than a fully assembled tether climber. Sufficient storage will be provided for multiple payloads, spare parts, and other materials.

The storage requirement is subject to all the same issues as the logistics requirement. The items to be stored will have unpredictable and potentially hazardous characteristics. Safety standards for the storage of hazardous items such as fuel and assembled rocket motors must be met. Some items to be stored, such as food products, will require temperature-controlled storage facilities.

3.9 Positioning

Earth Ports will be floating structures and will require positioning capabilities.

There are two primary alternatives for this capability. First, the Earth Port may be kept in place by thrusters of the type used to maneuver ships. Second, the Earth Port may be anchored to the sea floor and be equipped with powered reels on the anchor cables. In the second case, the Earth Port may still be equipped with thrusters, and positioning may be accomplished by a combination of thruster and reel operations.

The requirement for lateral positioning will be expressed in terms of acceleration and speed. These terms, plus others such as the mass of the Earth Port and its coefficient of drag while moving through the water, will be used to derive the power rating of the positioning subsystems.

3.10 Power

The Earth Port must generate and distribute power to all the other subsystems. For practical reasons, each platform will likely require its own power subsystem.

We estimate that a single FOP will require several megawatts of power-generation capacity. To achieve high reliability, we expect this capacity should be spread among multiple generators with one or two backups (N+1 or N+2 redundancy). Generators will likely be of the gas turbine type and fueled by natural gas, propane or liquid fuels of the same type used for OGVs and aircraft.

The largest power demand will be seen during the first 40 km of the ascent of the integrated tether climber and shroud. This power will probably be transmitted through a power cord. Only one climber at a time will be powered this way. We estimate this figure at about 4 MW.

There are some who believe that the Earth Port may also provide ground-based laser power transmission to tether climbers. This method of power delivery can be used with multiple tether climbers at the same time, so additional power generation capacity may be required. Also, laser power may be provided from separate floating platforms.

3.11 Water

It is preferable that fresh water be produced from sea water via desalinization. This is, however, an expensive process and the possibility of transporting fresh water from shore should be considered as well. The fresh water must be stored and distributed within the Earth Port.

Gray and black water must be stored and processed appropriately before being released or otherwise departing the Earth Port.

Sea water may be used directly in firefighting operations and for other purposes.

3.12 Maintenance facilities

Keeping the Earth Port operating will require maintenance facilities including a machine shop, paint shop, electronics shop, etc.

3.13 Safety

Ensuring the safety of the Earth Port requires attention to multiple requirements.

Physical security will involve equipment for surveillance and active threat mitigation operated primarily by dedicated personnel.

Environmental safety comprises facilities for observing and mitigating fire. Firefighting equipment including a water supply, hoses, nozzles, and so on, will be operated by both dedicated and part-time personnel. Some level of operational capability will be provided in the absence of facility power, perhaps by emergency power generators, pressurized water reservoirs, or other means.

Issues related to meteorological, oceanographic and man-made hazards, which may include the unintended release of pollutants from the Earth Port itself or nearby vessels, will also be monitored.

Meteorology is of special concern to the Earth Port because of the sensitivity of the tether and tether climber to weather conditions while moving through the atmosphere. The Earth Port will have its own specialized meteorological equipment and personnel, and will also receive and process meteorological information from other sources.

3.14 Security

The FOP will be painted in red and white checkerboard to enhance visibility. Its position will be reported to oceanic and space centers to diminish the probability of collisions. It will have audible and visible warning signals and a keep-out zone for safety and security. It will have defensive capabilities and operations staff will augment security personnel when needed.

3.15 Communications

Each floating platform will have equipment and personnel to maintain and operate communications channels to other platforms, other nodes in the space elevator system, tether climbers, free-orbiting satellites, ships, and so on. Its position will be reported to oceanic and space centers to diminish the probability of collisions. In addition, the location and activity of the Marine Node will be maintained and broadcast to all aeronautical organizations to ensure safety of flight near the space elevator.

3.16 Offices

Earth Port and customer personnel will be provided with suitable offices and other work areas. These will be equipped with computing and communications equipment as needed.

3.17 Quarters

Personnel on the Earth Port will also be provided with suitable quarters, kitchen/dining facilities, recreational facilities, shower facilities and so on.

4. Earth Port Description

4.1 Introduction

The Earth Port (formerly the Marine Node) Segment Definition and Mission, given in ISEC Position Paper #2014-1 states, in part, the following:

> *Basically, the Marine Node provides a location for the tether terminus that can enable safe and routine operations. Its primary purpose is the mating and de-mating of satellites and climbers. This would include stabilizing the tether, moving the tether, loading and unloading cargo, and local operations support. It is where the climber is prepared and then "sent on its way" safely. This fundamental support of the two-way transportation of goods connects the Space Elevator to the Rest of the World and the Rest of the World to the Space Elevator. The Marine Node will tie together all of the aspects of the terrestrial component to include safety, security, inspection of cargo, loading of cargo to climber, loading climber on the space elevator tether, off-loading climbers, and support to teams in the area.*

> *The Marine Node is a city on multiple floating platforms in the eastern Pacific Ocean. The main element of the Marine Node, the Floating Operations Platform (FOP), will be the size of an aircraft carrier or larger. Secondary elements are now envisioned outlying the FOP to give the tether terminus a strong base leg for tether anchoring stability. The Earth Port will have living quarters, kitchens and laundries, as well as recreational and medical facilities. It facilitates helicopter landings, local support watercraft, and the loading/unloading from larger Ocean Going Vessels.*

> *The FOP hosts a local Operations Center for the management of tether, tether terminus, and platform operations. In addition, the Center supports climber operations, including the operations and maintenance of the tether. The FOP vision is just now growing into something specific.*

This section covers the work by the Study Team to expand on the above statement and to provide a vision for the Earth Port and its primary elements that is indeed "something specific." This Section has also been developed in view of the functional requirements described in Section 3 and serves as the basis for the Staffing and Organization, and "Day in the Life" content.

The description of the Earth Port and its operations can be logically broken down into three related elements:

- transportation of cargo and personnel (including the OGV, local service watercraft, helicopters and other aircraft),
- the Floating Operations Platform (the virtual "city at sea") and

- the tether terminus platform(s).

4.2   Ocean-going vessels, service watercraft and aircraft fleet

4.2.1   Transportation considerations

The Space Elevator system can be visualized as a two-way transportation system that allows cargo to be moved efficiently from a point of origin to the GEO node and beyond. It will also allow such potential enterprises as the mining of the Moon or Mars for valuable minerals and safely returning them for processing and/or distribution on Earth. Other previously stated goals of a space elevator include space tourism and the facilitating of colonies on other bodies within the Solar System.   This transportation system, from payload origin to GEO and beyond, is conceptualized in Figure 4-1.

As indicated, the origin or destination of space elevator cargo can be virtually anywhere on Earth that has reasonable access via ground, sea, or air to a city that has an international airport and/or commercial harbor. From this point, the cargo will be transferred to a selected Earth Port access city by air or cargo (container) ship. The access city should be within a feasible distance from the remotely located Earth Port so that personnel and payload can be safely, efficiently and comfortably transported by high speed ferry, ocean-going tug/barge combination, smaller general cargo ship or long-range seaplane. These OGVs and aircraft are more fully described in Section 4.2.2 below.

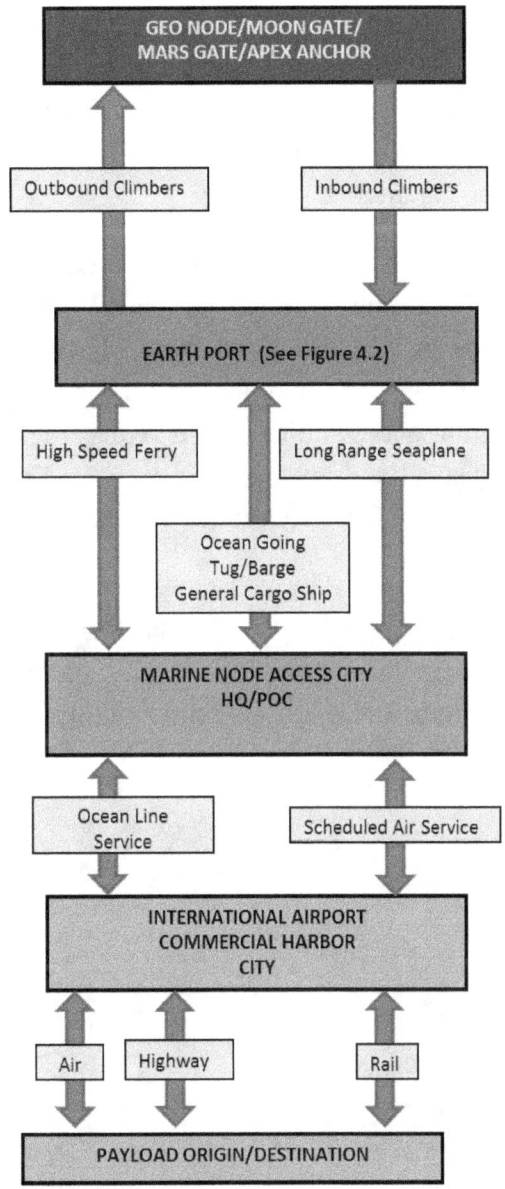

Figure 4-1 Transportation System

.

In ISEC literature and studies to date, the preferred location for the Earth Port is in the eastern Pacific Ocean, on the equator, some 1,000 kilometers west of the Galapagos Islands. This location is about 1,300 air miles from Quito's international airport in Ecuador. Ecuador's three small commercial harbors, Guayaquil, Esmeraldas and Machala lie approximately 1230 nautical miles from the "preferred" Earth Port's location. For reference, the "preferred" Earth Port location is about 1,350 air and nautical miles from Panama City and the Port of Balboa in Panama. It is a little more than 1,000 air and nautical miles from Acapulco and over 2,100 miles from San Diego's international airport and deep water commercial port.

In determining the final location for the initial Earth Port facilities, a number of factors will come into play. These include: international laws associated with sea, air and space, year-round weather and wave conditions, underwater topography and access to emergency services. Serious consideration should also be given to the location of the Earth Port access city. This city, or cities, should be served by regularly scheduled international airlines and cargo shipping liner services. The safety and reliability of the payload transportation link, and the client personnel associated with it, should be paramount.

Another significant factor to consider in locating the Earth Port access city is the political stability of the country in which it lies. A review of major port/airport cities on the Pacific Ocean within a reasonable distance of equatorial waters, and having a stable political regime, would bring into consideration such cities as Honolulu and Singapore. Keeping short the travel distance and time from these potential access cities would require determining a new location for the Earth Port. For several logistical and business reasons, the ultimate Earth Port access city will likely be the location for the Head Quarters/Primary Operations Center (HQ/POC) facilities.

4.2.2 Ocean-going vessels and aircraft fleet

Due to its remote location in the eastern Pacific, the Earth Port will be serviced by a fleet of OGVs, work boats and aircraft that are either owned or leased. A view of the transportation operations to/from and within the Earth Port, is shown in Figure 4-2.

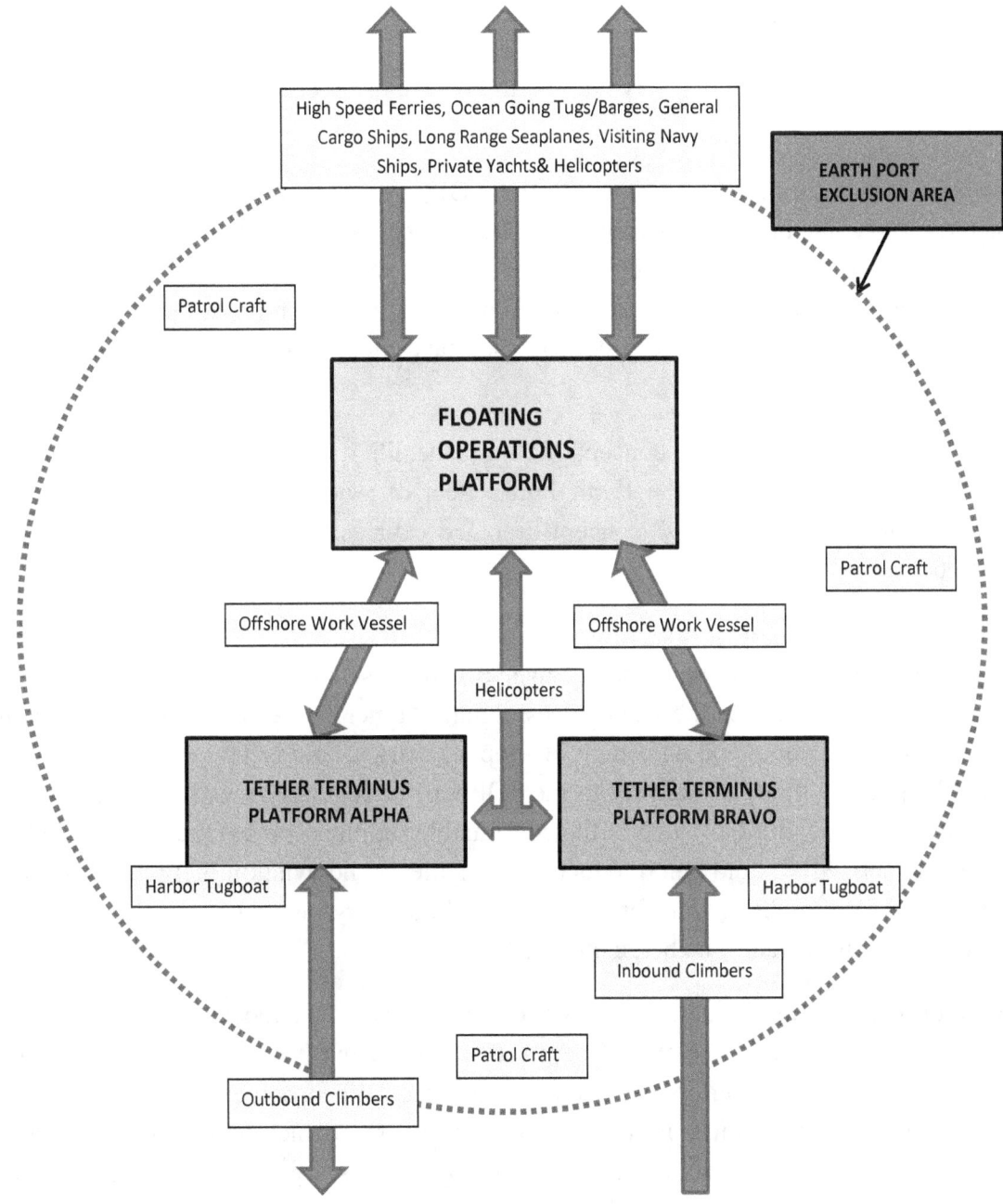

Figure 4-2 OGV and aircraft

In the descriptions presented here, it is presumed that the various cargo to be lifted into space by the tether climbers of the Earth Port will be some combination of: high value, complex, bulky, fragile and secret. As such, these payloads will most likely be placed in a secured standard marine shipping container (8 ft. by 8 ft. by 45 ft.) at some point prior to departing the Earth Port access city. It is further presumed that the shipments of climber payload to the Earth Port would occupy only a few containers at any given time. Because the Earth Port is remote from normal supply channels, the OGV fleet will also provide necessary supplies for the long-term operations of the Earth Port platforms.

The sea and air vehicles that will serve the Earth Port are discussed in detail in Appendix C.

4.3 Floating Operations Platform

In various ISEC documents to date[4,5,6], the FOP, the core of the Earth Port, has been characterized as something "larger than an aircraft carrier," a modified offshore drilling platform, a converted ocean-going drill ship, or a "city at sea." Common to all these depictions is the need for the platform to maintain station in the open ocean environment on or near the equator in the eastern Pacific and to have all the necessary facilities to carry out the functional requirements described in Section 3 above.

This section will address these concepts in more detail. It will also address the "on-board" facilities necessary to support the Earth Port's fleet of watercraft and aircraft and its cargo handling capabilities. A new FOP concept designed expressly for the Space Elevator program will also be presented.

4.3.1 Cargo handling, storage and transport facilities

The remote location of the FOP in the open ocean presents several design and operational challenges. Among these is the need for essentially "quiet" water (less than 1 foot waves or surge) for the safe loading and unloading of supplies, cargo, liquids, containerized payloads and people. The primary mode of delivery to the FOP will be regularly scheduled ocean-going tugs and barges (see Appendix C). Much of the consumable supplies (stores) received at the Platform will be containerized. Liquid bulk products such as diesel and aviation fuel will arrive in barge-mounted cylindrical tanks or transferred from small auxiliary general cargo ships. Pressurized gases such as butane can also be barge mounted.

In order to achieve quiet water in an open ocean environment, the FOP itself must act as a breakwater. As such, it must be able to be position itself using its own thrusters or engines to a proper alignment based upon oceanographic predictions. When barges or ferries are alongside the loading area of the platform, it may be also be prudent to deploy a floating breakwater using patrol craft and/or outboard utility boats. Figure 4-3 shows a proprietary floating breakwater system that can attenuate up to five-foot waves down to acceptable heights. U.S. firms such as Whisprwave™ and WaveBrake™ can design and manufacture the breakwater barriers.

---

[4] Ibid.

[5] R. Penny, P. Swan and C. Swan, Space Elevator Concept of Operations, ISEC Position Paper 2012-1, 2013.

[6] P. Swan, C. Swan, R. Penny, J. Knapman and P. Glaskowsky, Design Considerations for Space Elevator Tether Climbers, ISEC Position Paper 2013-1, 2014.

Figure 4-3   Floating wave attenuator system

There will be breakwaters to meet operational requirements for safe cargo loading/unloading and personnel boarding.

In addition to having quiet water for cargo loading and unloading, the FOP must have a suitable work area with an elevation 17-20 feet above local sea level. At this elevation, cargo and passengers can be easily and safely transferred from the deck of the barge or ferry as well as the passenger doors on seaplanes.

As much of the cargo will be containerized, it would be appropriate to have a wide, 100 foot or so, clear working area for interim storage, similar in concept but smaller in size, to a typical and based container terminal layout. Rubber-tired mobile crane(s) with sufficient outreach can then work the dock area unloading inbound containers from barges and loading containers bound for tether terminus platforms onto offshore service vessels. An example of such a device is shown in Figure 4-4.

Figure 4-4 Rubber-tired mobile container crane

Once a container is on the work deck of the FOP, it can be moved into an interim storage area, supply storage "locker", refrigerated space, below deck access hatch or workshops by a heavy-duty, rubber-tired fork lift similar to the equipment shown in Figure 4-5. The FOP should have a variety of storage facilities for liquid, dry and refrigerated cargo that serves the day-to-day needs of the personnel aboard.

Figure 4-5 Heavy-duty forklift

It should also have a number of assigned work areas for client payload assembly and preparation, including so-called "white rooms." As an example, once a payload has been prepared by client staff, it can be re-containerized, moved to the loading deck by forklift and then loaded onto an offshore service vessel by the mobile crane.

4.3.2 Fleet service and maintenance facilities

The FOP, as a semi-submersible structure in the remote open ocean, has a somewhat unique role to play when compared to naval or commercial ships, off-shore oil drill platforms or floating storage facilities (converted tankers.) That is, it must be a full-service marina for the small fleet of watercraft described in Section 4.2.2 above. As such, the FOP must provide protection for these craft from adverse atmospheric and oceanographic conditions. The FOP must also have facilities for refueling, resupply and maintenance, including spare parts.

All of the various types of platforms envisioned by previous ISEC studies are relatively massive and have self-propelled, station-keeping capabilities. If properly oriented to oncoming wave energies, they can provide a "lee" area adjacent to the platform for mooring of the smaller vessels when not in use. It would also be prudent to provide cover for these vessels at an elevation higher above local sea level than the largest air draft in the small fleet. This would reduce maintenance requirements resulting from exposure to rain and salt-laden sea air. As any small boat owner or former Navy deck hand knows, well-planned maintenance activities are a must.

To accomplish this required maintenance, the FOP will have a number of specialized shops and repair stations, including: a machine shop with engine overhaul capabilities; a battery repair and replacement shop; paint "lockers"; parts storage for all of the small craft; and electronics service and repair.

Readily accessible refueling stations for various products would also be required. In the case of the seaplanes and high-speed ferries, refueling pump stations should be at or near the cargo and personnel off-loading work areas.

The FOP should also have at least two designated helicopter landing areas on its top deck with appropriate refueling stations, maintenance shops and storm or security tie-down capabilities. In addition to helicopters, the Earth Port will probably have drone-type aircraft for patrol purposes. The FOP must also provide secure storage and maintenance facilities for these sensitive aircraft.

All of these facilities would be in addition to those required to operate and maintain the FOP and Tether Terminus Platforms described in Section 4.4 below.

4.3.3 On-board weather forecasting and oceanographic facilities

4.3.3.1 Meteorological equipment and facilities

The potential impacts of adverse weather conditions (wind, precipitation, and lightning) on the tethers and climbers from the surface of the ocean to, at least, 40 kilometers are of paramount concern to the day-to-day operations of the Earth Port. As such, the FOP will be linked continuously to the HQ/POC. The POC will use geostationary and low-earth-orbit weather monitoring satellites for the sensor data on and around the FOP to generate the products needed for operations. The most important decisions are the "GO" and "NO GO" decisions. Analyses of this data will include both macro and local climate predictions that will factor into the operational planning of launch and retrieval operations. There is a wealth of experience gained through the launching of rocket-powered satellites over the years that will be applied, as appropriate, to the tether/climber operations. The parameters of concern are:

- lightning onset, cessation, detection and warning reports,
- wind in the vicinity of the FOP as well as through the "column" to 40 km,
- visibility due to clouds and mist/haze,
- thunderstorms (particularly in the vicinity of the platforms) and
- moisture content of nearby clouds.

Aboard the FOP, there will be a local weather station for collecting and analyzing temperature, humidity, wind, air pressure, wave height, and precipitation data. The FOP should have facilities for launching and retrieving weather instrumentation packages by balloon. Due to the marine environment and the potential for corrosion in delicate instruments, there should be a maintenance and repair shop on-board. Upper air measurement systems may include RainWindsones, Doppler radar wind profilers and sensors on the climbers and at regular intervals on the tethers up to at least 40 kilometers. Associated with this local and immediate meteorological data collection will be a state-of-the-art array of weather sensors and communication antennae, probably on the top deck of the FOP. Typical images of this equipment are shown in Figures 4-6 and 4-7.

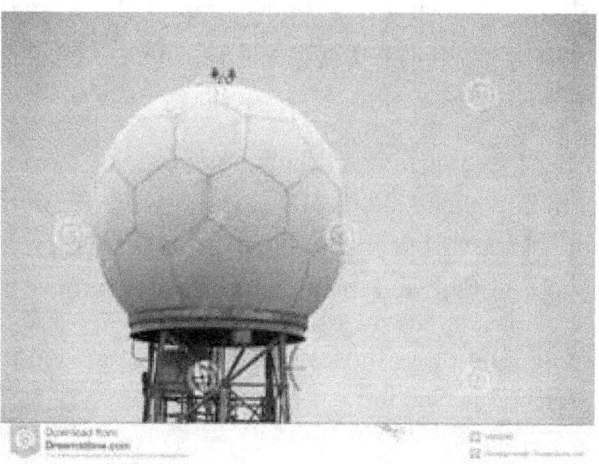

Figure 4-6   Doppler radar dome

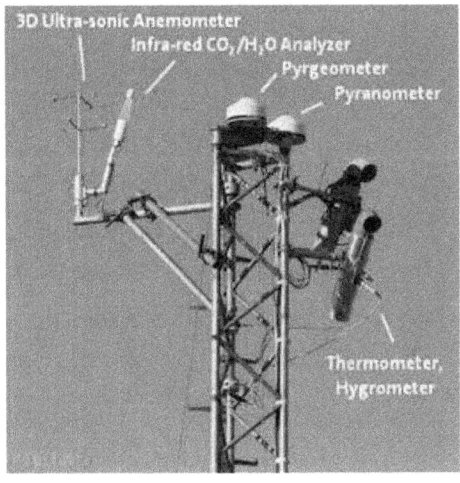

Figure 4-7 Meteorological instrumentation

Within the FOP Operations and Control Center there will be stations for data ingest and display as well as local numerical prediction model(s.) Communications between the FOP and the tether terminus platforms will also be facilitated from the operations center.

4.3.3.2 Oceanographic instruments and facilities

In addition to the dynamic advantages of operating the Space Elevator system from the Equator, there are meteorological and oceanographic advantages as well. The equatorial Pacific Ocean lies in the so-called "doldrums" or the Inter-tropical Convergence Zone. As stated by Gardner (2003) the preferred location for the Earth Port west of the Galapagos Islands that lies within this Zone had "....one lightning strike per year per square kilometer, very low probability of hurricanes and cyclones and almost no wave issues."

In order to assure safe operations for OGV cargo loading/unloading, ferry and/or seaplane berthing and protection of the Earth Port's fleet at the FOP's open ocean location, wave climate must be predictable and continuously monitored using a variety of oceanographic systems and equipment. Scientists at the HQ/POC and oceanographic technicians at the FOP can readily "plug into" the wave databases developed and maintained by NASA, the National Oceanic and Atmospheric Administration (NOAA) and its National Centers for Environmental Prediction (NCEP).

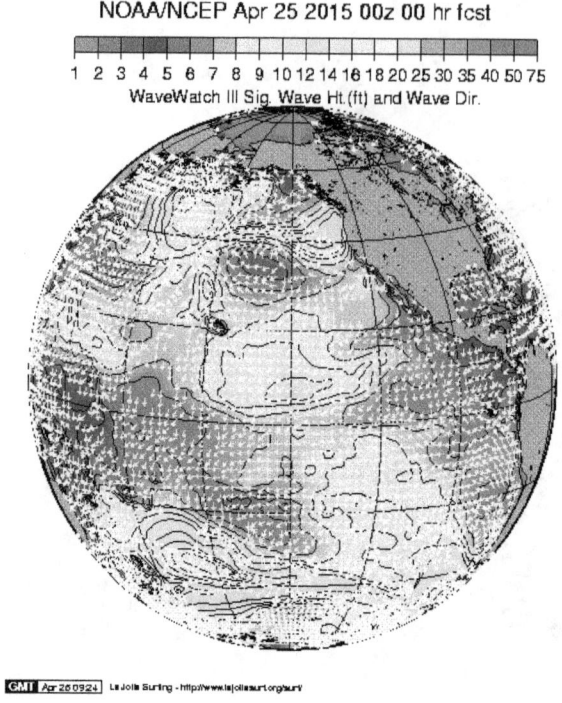

NOAA/NCEP Apr 25 2015 00z 00 hr fcst

1 2 3 4 5 6 7 8 9 10 12 14 16 18 20 25 30 35 40 50 75
WaveWatch III Sig. Wave Ht.(ft) and Wave Dir.

Figure 4-8   WAVEWATCH III data

Historical baseline data including wave height, period and direction, from NOAA's National Data Buoy Center can be obtained and used in discussions about candidate sites. Current macro wave conditions at the Earth Port's location can be obtained from the NOAA/NCEP WAVEWATCH III system. A sample of this data, for wave height (ft) and wave direction on April 25th, 2015 for the globe, is given in Figure 4-8.

In the immediate vicinity of the Earth Port, moored or floating surface buoys can be placed at calculated distances from the platforms with instruments for continuous monitoring of local wave spectra. Analyses of this data at the FOP's operation and control center will be used to direct the orientation of the FOP and tether terminus platforms to the prevailing wave energy. From this data, the need to deploy the floating wave attenuators will be determined.

Figure 4-9 FOP operations and control center

The FOP hosts a local operations center for management of tether, tether terminus, and platform operations. In addition, the center supports climber operations including operations/maintenance of the tether. Similar to the bridge of a marine vessel, the Operations and Control Center (OCC) will be occupied by one or more watch officers on a continuing basis. Its communication linkages to/from the HQ/POC and operations aboard the tether terminus platforms at any given time will be of paramount importance. A notional center is shown in Figure 4-9.

One of the principal features of the OCC will be status displays of the various in-transit climbers to and from the GEO node. Similar to conventional space launch operations, there will be space, equipment and displays for essential client personnel as well as an observation area for client and visiting V.I.Ps.

As mentioned above, data display and analyses from meteorological and oceanographic instruments will probably be located in the OCC. Reports from the various functional departments aboard the FOP will also funnel into the OCC via computer displays, communications headsets, virtual reality systems and direct contact.

Security of the Earth Port is a continuous operation. Deployment of high-speed patrol water and aircraft and/or drones will be directed from the OCC. Long-range underwater sonar readings will also be monitored and reacted to at the OCC. In addition, during emergencies, the OCC will probably act as the Emergency Control Center for the entire Earth Port.

As described in Section 4.3.5 below, the Engineering Department may have a station in the OCC that monitors all of the on-board systems that fall within that department's responsibilities.

As the OCC is developed and staffed, the latest advances in computers, electronic displays, software programs and communication devices will be incorporated, as appropriate.

4.3.5 Engineering facilities

The Engineering Department or Division aboard the FOP will have a wide range of behind-the-scenes responsibilities that will keep the "city at sea" functioning around the clock. In many respects, the engineering "plant" on the FOP will be similar to that of a large U.S. Navy combatant or auxiliary ship. The plant may also be similar to that of a large container, cruise or drill ship. There is one significant difference however: the FOP will be on permanent station in the open seas without the opportunity to enter ports from time to time to resupply its stores and maintain its infrastructure. In other words, it has to be self-reliant and very efficient.

4.3.5.1 Primary systems

Most of the primary systems operated and maintained by the Engineering Department will be "below decks" no matter what final size and shape the FOP takes. Final design of these facilities and systems will be the province of marine architects and structural engineers. Major systems will include the following:

- ballast water tanks with pumps and piping to maintain vertical stability of the overall platform structure in the open sea,
- fuel tanks of various types with pumps and piping storing barge delivered liquid and gas products used for generating main power, supplying on-board equipment and refueling the Earth Port fleet of watercraft and aircraft,
- main propulsion engines and horizontal thrusters for station-keeping and strategic orientation of the platform relative to local wave energy spectra,
- sea water desalinization plant, potable water storage and distribution system to spaces throughout the platform vessel,
- sewage collection, treatment and discharge (into the ocean) equipment including automated waste collection system from throughout platform compartments,
- solid waste recycling system possibly including some energy production from waste materials,
- fire suppression sea water intake pumps, piping system and related equipment stations,
- air conditioning equipment, heating and ventilating equipment and ducting,
- refrigerated storage equipment and
- cargo and personnel elevators from working cargo handling deck to upper/from upper working decks.

As a virtually independent, multi-functional facility with locational constraints and limited resources, the FOP, by necessity, must be efficient in the highest sense of the word. In order to achieve this overall efficiency, the developers of the FOP should apply the best available "smart" technologies to all of the systems mentioned above and below. Using advances in digital information technology, sensors and monitoring devices, data sharing software and systems integration hardware, all of the engineering equipment can be integrated into a controllable dynamic network. Integrated smart devices, as simple as space lighting sensors, can reduce power consumption loads by turning off or dimming lights when spaces are unoccupied. The detailed design of the Earth Port's platforms will be developed in parallel with tether materials research, climber design, solar power panel design and manufacture, robotics for the GEO Node, etc. Work being done in the fields of city planning, civil engineering and information systems technology toward the development of "smart cities" and smaller urban areas will provide valuable insights.

4.3.5.2 Electrical power generation and distribution system

An analysis of the overall electrical power for the FOP has determined that an 11 to 12 megawatt (MW) generating capacity will be required at full operations. There are a number of alternative methods available today and in the near future to generate the power required for the operations of the FOP and tether terminus platforms. Solar, wind and ocean wave energies can be converted to electrical power and would all be available to varying degrees at the Earth Port's location. Several researchers and manufacturers are developing small nuclear powered engines that would be of a size suitable for the FOP's needs.

However, the most cost-effective approach would be to utilize the mature technologies of high-power gas turbine generators. Their overall dimensions, weight and cost seem to be suitable for the FOP's needs. Shown in Figure 4-10 is a 14.25 MW natural gas generator that is 15.5 by 3 by 3 meters in size and has a weight of 72 tons. A likely scenario would be to have two of these running with a third on standby or in maintenance.

Figure 4-10 Gas-powered generator set

Associated with the generators would be a number of electrical distribution panels, step up and step down transformers, control room(s), cabling, insulation and duct work.

4.3.6 Hotel facilities and visitor accommodations

Section 5 describes the organization and staffing requirements for the Earth Port. Facilities that provide living accommodations, food service, health care, shop, storage/office space, education, and recreation will occupy much of the middle levels of the FOP. Many of the personnel aboard the FOP will be serving for extended periods of time, say six- to nine-month "rotations." This reality adds to the size and number of staff support accommodations required when comparing the platform to a cruise ship or navy vessel.

The tropical ocean environment and relatively benign weather associated with the location of the Earth Port provides opportunities for "quality of life" enhancements aboard the FOP. One can envision off-duty staff members water-skiing behind one of the high-speed utility boats, sailing in small yachts, fishing, playing tennis, basketball or volleyball on one of the open decks or even driving (biodegradable) golf balls into the Pacific. Skeet shooting may even be allowed from a deck overlooking the ocean.

Staff quarters and staterooms would probably range in size from 200 to 350 square feet, similar to those on a typical international cruise ship. These quarters would probably be individual and "smart."

Normal daily food service will be provided in a central dining room supported by efficient food preparation, scullery, waste disposal and refrigerated storage areas. Coffee, tea and snack (pizza) stations and/or vending machines may also be located in key areas around the platform. Depending on the standards of the company (ies) managing the platform, a pub space may also be provided for social and recreational activities aboard. Clients and other visitors may wish to have food service in a high-end, restaurant-style facility.

Associated with the food preparation and serving functions there should be space aboard for growing fresh vegetables and herbs in soil beds, hydroponically and/or aeroponically.

The FOP will probably have multi-purpose rooms that can serve as a general meeting hall, library, large-screen video theatre, card room, etc. A general store will also be provided for staff and visitor retail needs while aboard.

A satellite communications center should be provided for staff and visitors to keep them in touch with their respective homes and offices.

A pharmaceutical dispensary, emergency operating room, hospital rooms and related medical equipment and facilities will be also provided aboard in the platform's "sick bay" area.

Clients and visitors will be provided temporary staterooms, offices and conference facilities aboard. At full operation capability, there may be up to 100 or so client visitors on board at any given time.

Based on the above discussion, the space required aboard for crew and visitor serving facilities would be in the range of 10,000 square feet.

4.3.7 Conversion of existing marine "platforms" to FOPs

In ISEC studies to date, a number of existing marine vessels or platforms have been mentioned as candidates to serve as the FOP. A sampling of these is presented below along with images and an approximate displacement mass expressed in terms of metric tons. One metric ton equals 2,204 pounds.

Nimitz **Class Aircraft Carrier**

1,092 feet LOA by 252 feet beam by 41 foot draft. 3,200 ship crew members plus air wings.

80,600 metric tons. nuclear propulsion. Unlimited range.

Figure 4-11 USS Nimitz underway

Rowan Drill Ship

752 feet LOA by 118 beam by 59.6 foot draft.

Crew accommodations: 210.

69,900 metric tons.

Diesel engine main propulsion with 4 retractable horizontal thrusters.

Figure 4-12 Rowan Reliance drill ship

Figure 4-13 Sedco 704 drilling platform

Sedco D704 Drilling Platform

(300 feet by 263 by 70 foot draft. Crew accommodations: 117.

23,700 metric tons

Diesel engines main propulsion with 4 electric powered horizontal thrusters

May also be towed or transported into position

If the launching and retrieving of payload climbers and tether handling is to be accomplished on independent platforms as described in Section 4.4, then other types of existing ships may be considered for conversion. As a prime example, the physical requirements for crew, client and visitor accommodations described above would make a small cruise ship a likely candidate.

Existing drill ships and platforms meet the FOP's requirements for open ocean stability, station-keeping, cargo handling and helicopter operations. They seem to fall short in their respective capacities for personnel accommodations and supporting the needs of the Earth Port's small fleet of service craft. A U.S. Navy (surplus) aircraft carrier may be too large in terms of the crew size and costs necessary for its continuing operations.

In this study, consideration has been given to designing and constructing a FOP vessel to expressly meet the requirements of the Earth Port. One concept for this "new construction" is discussed in the following Section.

4.3.8. FOP new construction concept

Figure 4-14 FOP concept

This concept will evolve and be addressed in future reports.

4.3.8.1 Below deck facilities

The design of the twin hulls shown in the conceptual image will be determined by naval architects and structural engineers. They may take the shape of a "standard" ship's hull or that of a semi-submersible offshore structure. In concept, the two hulls will be tied together at the cargo deck level and underwater at a depth below the draft of any of the vessels that will be moored under the main deck canopy or alongside the cargo loading deck. For comparison to the potential conversion platforms, the dimensions would be in the range of: 600 feet long by 400 feet breadth by 80 feet draft with a mass between 50,000 and 60,000 metric tons.

The FOP may have its own main propulsion system or be towed or transported to the Earth Port location using heavy-duty, ocean-going tugs or special transporter ships as shown in Figure 4-15.

Figure 4-15 Alternative methods to move platform to Earth Port location

In any event, the FOP will be equipped with electric motor power horizontal thrusters to maintain station and orientation to wave energy.

The interior space within the hulls of the FOP will be segmented into decks and tanks that will house or contain the following:

- main power plant and propulsion system,
- electrical generation and distribution system,
- main refrigeration and air conditioning plant,
- diesel fuel, aviation fuel, liquefied gas storage tanks,
- desalinization plant and potable water storage,
- liquid waste treatment plant and solid waste center,
- sea water intakes and fire suppression pumps,
- horizontal thruster machinery and
- ballast tank system.

4.3.8.2 Cargo deck level and covered berthing area facilities

Two essential functions of the FOP described in Sections 4.3.1 and 4.3.2 are the primary drivers in determining the shape and size of the conceptual platform described. These are:

- providing an open cargo handling deck for containerized supplies and client payloads at 17-20 feet above local sea level and
- providing a protected area at sea for the berthing and servicing of the Earth Port's fleet of watercraft.

Tug and barge mooring, with appropriate fendering system, would occur along the "east" face of the FOP. A floating wave attenuation system may be deployed to assure quiet water during barge loading/unloading operations. See Figure 4-16 below.

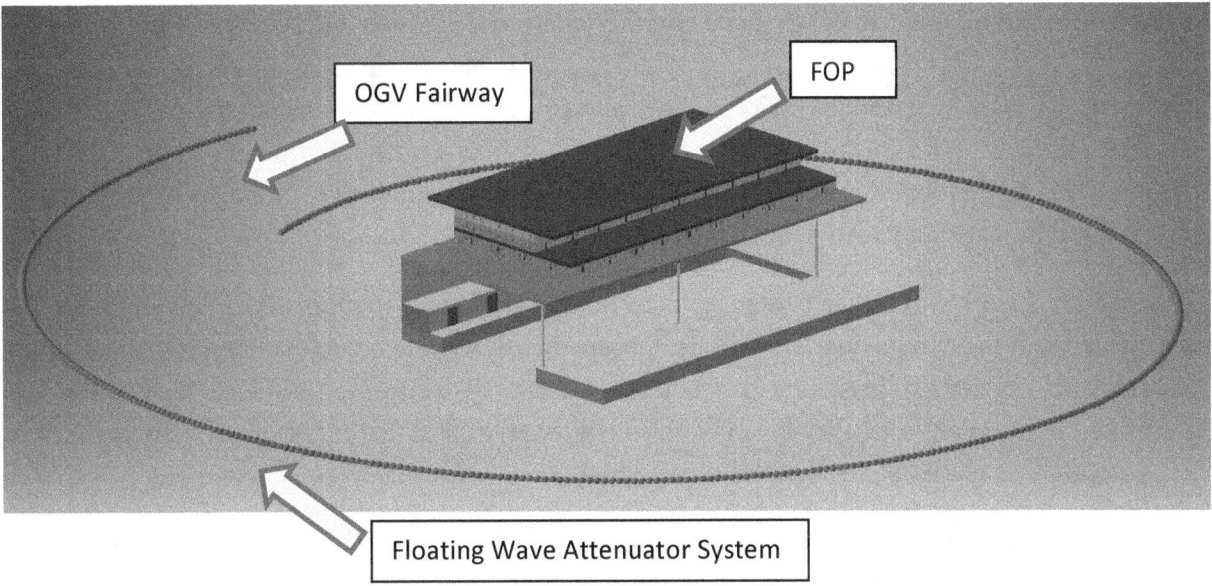

Figure 4-16 Floating wave attenuation system

The length (minimum 600 feet) and width (minimum 150 feet) of this deck will be adequate for the moving and interim storage of containers using mobile crane(s) and heavy-duty forklifts.

Along the far wall of the platform will be number of large 75 feet wide by 50 feet high, 75 feet deep bays as shown in Figure 4-14. There will also be storage rooms or open bays, refrigerated spaces, marine operations offices, specialized maintenance shops, elevator shaft(s) and stairwells to the other decks. Clients will be assigned secure spaces to open their respective containers and prepare their payloads for transport to the tether terminus platforms and climbers.

A clear working area around the internal small craft berthing area will be maintained for activities similar to those of a typical marina. As necessary, this area may also be used for loading and unloading of passengers and cargo from the high-speed ferries or long-range seaplanes servicing the Earth Port. A unique feature (versus a converted ship) of the proposed concept is to have this "marina" area covered by the underside of the main deck. The overall vertical clearance will be determined by the maximum air draft of the vessels to be berthed inside the structure (see Section 4.2.2).

4.3.8.3 Superstructure facilities

One of the structural challenges of the FOP concept is to support the main deck and above by columns along the edges of the cargo deck so that the approach to covered berthing is clear of obstructions and that there is sufficient open space for container handling operations on the cargo deck.

As conceptualized, the largest of the superstructure decks ("main deck') includes custom designed air-conditioned interior space for the tether and climber OCC, Earth Port command, business and administrative offices, communications center, central computer equipment area, conference rooms and client office space, meteorological and oceanographic center, main engineering control, Earth Port security center and other essential functions. The open space on the main and upper decks may be used for outdoor recreational activities including deck sports. A safe smoking area may also be provided.

The mid-level deck will be the center for crew support activities mentioned in Section 4.3.6. It is anticipated that the crew's quarters will be equipped with the latest personal communications and entertainment systems. There may be sufficient outside space on the mid-deck for creation of a park-like environment for off-duty crew and visitors as well as garden space for the growing of fresh vegetables and herbs for the platform's chefs.

The top deck will be an open and bustling working deck that provides space and protection for: electronic communications antennae, radar antennae, weather station equipment including balloon launching and retrieval, helipads (possibly cantilevered) including tie-downs, and drone operations and maintenance.

4.4   Tether Terminus Platform(s)

4.4.1 Platform description

The floating platform image shown in Figure 4-17 has been used in ISEC study reports, journals, posters and other publications for several years. It is a good portrayal of the tether terminus platforms.

Figure 4-17 Tether terminus platform [image by Chase Studios]

The figure also shows in the foreground one of the tethers with a payload climber underway (outbound) and in the distant background, a companion (inbound) platform and tether. This platform would be similar in dimensions to the *Sedco 704* semi-submersible drilling platform shown in Figure 4-13, about 300 feet by 275 feet by 70 feet draft at 25,000 metric tons. The platform shown has several features that make it suitable for tether launch and retrieval operations. An offshore service vessel is alongside a low-level platform discharging payload cargo. A fixed crane lifts the cargo to the working deck. Climber crews and client personnel arrive and depart via the short-range helicopter shown on the helipad. Communications with the FOP's operations center and HQ/POC are maintained via satellite up/downlink antennae. Housings for the tether reel and power supply reels can be seen as well as the climber protective equipment and on-board power generator. A harbor tug could be also shown alongside to augment the platform's thrusters in rapid response movements that may be required to avoid space debris. To complete the picture, one of the high-speed patrol boats could be shown circling the platform.

Much of the engineering equipment and systems aboard the FOP will also be required aboard the tether terminus platforms, particularly the ballast system to maintain their stability during daily climber operations. There would also be an enclosed air-conditioned space for the platform's respective communications and engineering control equipment and monitors. Workshops and

33

stores for minor repairs of the on-board equipment would be necessary. Basic conveniences for the working crews aboard, such as sheltered break areas with coffee stations and vending machines, should be included. Automated perimeter security monitors linked to the FOP's Security Office would be also be needed.

Referring again to Figures 4-17 and 4-2, the final distance between the pair of tether platforms will be determined by the dynamics of tether/climber operations on the surface of the ocean within the Earth Port's exclusion area. The position of the FOP relative to the tether platforms in both distance and bearing will also be determined by calculation but may be less critical for day to day climber operations.

4.4.2. Tether power supply (to 40 kilometers)

A number of alternatives for powering the climbers through the atmosphere to at least 40 kilometers above the ocean have been discussed in previous ISEC papers and conference presentations. ISEC Position Paper #2013-1 "Design Considerations for Space Elevator Tether Climbers" summarized the thinking on this subject up to its publication date in the spring of 2014. Subsequent work by the Earth Port study team has narrowed the option to the concept of a 40km long "extension cord" from the tether terminus platform to the point in space where solar panels take over for the continuing climb to the GEO Node.

The power requirements for the initial climb, reel-in and reel-out capabilities, and thruster station-keeping were calculated in the current study to be about four megawatts. This level of power can be generated on board the tether platform using a commercial gas-powered generator. The so-called extension cord will possibly be a nano-conductor integrated with the tether material.

4.4.3 Onboard Climber Equipment and Facilities

Without repeating all of the information presented on the subject in ISEC Position Paper #2013-1, this section will mention the additional equipment that will be on board the tether platforms to carry out the primary function of launching and retrieving payloads to/from space.

At the FOP, the payload for a given climb has been received, temporarily stored, assembled, processed and configured as necessary. It will be transported over the ocean to the tether terminus platform by one of the offshore service vessels. As previously mentioned, it will be lifted by a fixed-platform crane to the working deck of the platform.

Aboard the platform, the climber, on a dolly, will be rolled to the proximity of the tether at which point the gripper mechanism is opened and the climber is properly positioned on the tether. The payload cargo is then loaded into the climber. This will require skilled use of a fork lift or mobile deck crane. The atmospheric protective shroud then receives the climber and payload. Figure 4-18 shows a conceptual rendering of this "mating" of loaded protective shroud with the tether.

Figure 4-18   Tether/shroud mating facility concept [image by Chase Studios]

At this point, the climber power extension cord is plugged in, final operational checks are made and the daily climbing operation begins. Section 2.5 "Life Cycle of a Climber" in the referenced ISEC Position Paper provides a clear narrative of the total climber operation involving the FOP, terminus platform equipment and transport vessels operating within the Earth Port.

4.5   High Stage One

The Earth Port has many variables that have been discussed over the years. One key concern for the operations is the first 40 kilometers straight up. Weather, winds, lightning, etc. are all hazardous to operations. One concept that has potential is to place a platform at that altitude and operate the tether terminus atop High Stage One. Appendix D covers this option as a quick summary. There are many reports and papers on the subject[7,8]. However, to be complete in this study, the concept of High Stage One is shown in a short version.

---

[7] Space Elevators: An Assessment of the Technological Feasibility and Way Forward, P. Swan, D. Raitt, C. Swan, R. Penny and J. Knapman, International Academy of Astronautics, October 2013.

[8] J. Knapman, "Benefits and Development of High Stage One," 63rd International Astronautical Congress, Naples, Italy, October 2012.

5. Earth Port Organization and Staffing

5.1 Introduction

The Floating Operations Platform, whether converted aircraft carrier, drill ship, offshore oil platform or new construction, will be staffed and operated as if it were a ship on a six- to nine-month deployment at sea. Specifically, it must remain on station in deep water on or near the equator at a significant distance from land, airports and seaports (as discussed in Section 4.2.1) supporting and carrying out the program's basic mission of launching payloads to, and returning them from, space using a tethered climber.

From an operational staffing viewpoint, it will be a "hybrid," lying somewhere in the spectrum ranging from a diesel engine powered container ship with minimal crew, a U.S. Navy Auxiliary ship, a modern cruise ship and a nuclear powered U.S. Navy aircraft carrier with thousands of personnel aboard.

Once Initial Operational Capacity (IOC) is achieved there will be on board, at any given time, three distinct groups of personnel:

- FOP command and crew,
- terminus platforms climber and tether operating staff
- clients, visitor and supply ship personnel.

These groups, their functional responsibilities and prospective "counts" are discussed in the Sections below.

5.2 Operating platform crew

Using a U.S. Navy aircraft carrier for reference, the crew of the FOP is the regular, steady force of personnel that "keeps the ship afloat" and functioning. This crew serves to host the aircraft squadron planes, equipment and personnel that operate from this floating base to carry out the primary missions of the ship. With respect to the Space Elevator, the primary mission is to launch and retrieve payloads to GEO and beyond via the tether climbers. In this sense, the FOP crew hosts the tether climber launch and retrieval personnel as well as visitors and supply ship personnel. In the following, it is assumed that this crew will stand four-hour watches twice a day along with their other technical duties, similar to a merchant or navy ship at sea. Staffing levels are predicated on regular rest times and appropriate days off. Figure 5-1 is a notional organization chart.

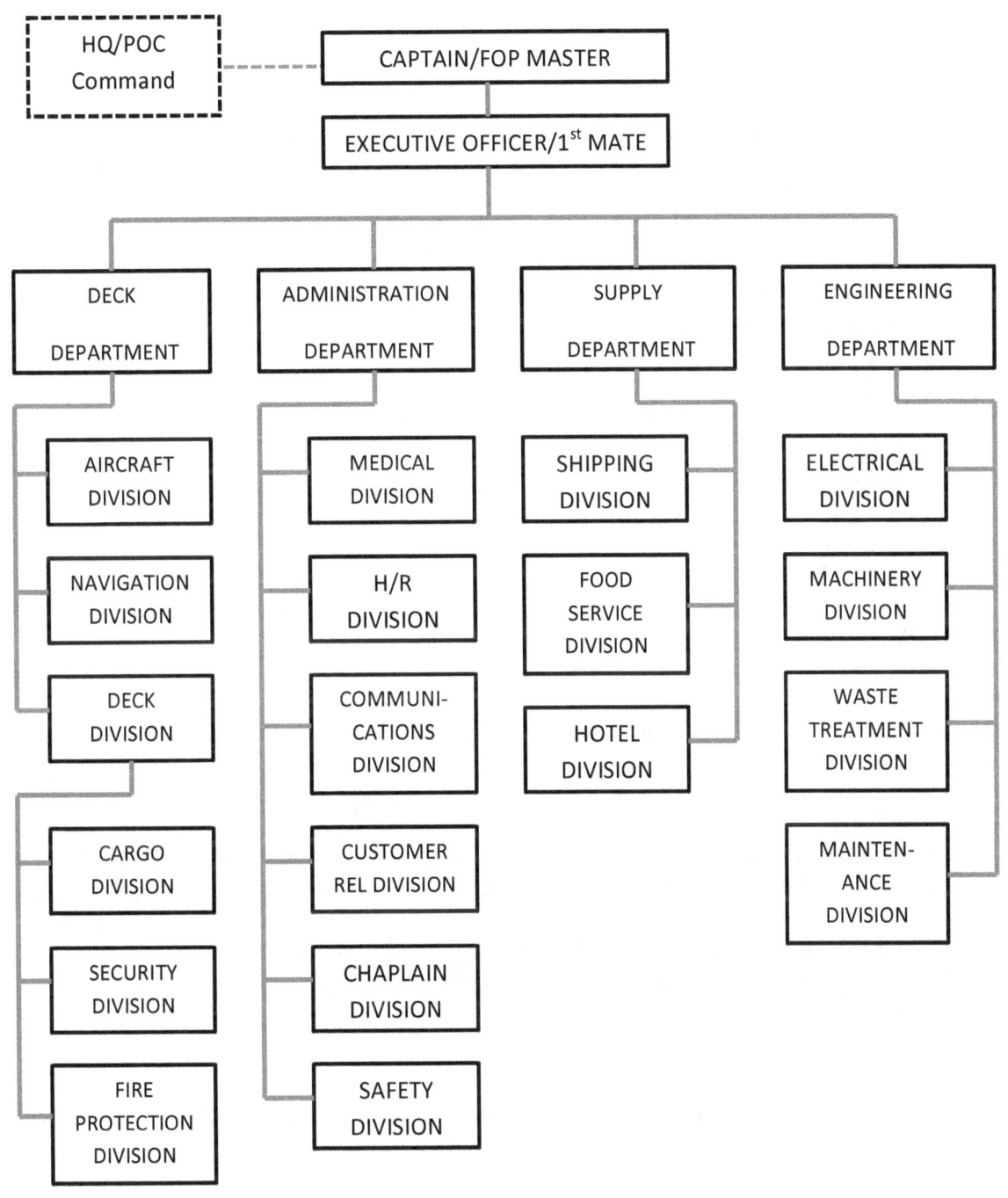

Figure 5-1 Notional organizational chart

5.2.1 Platform command structure

The Commanding Officer (CO) or Master of the FOP is responsible for all activities, facilities and personnel within the Earth Port. He/she reports to the CEO located at Space Elevator Headquarters (HQ/POC) that may be located at the Earth Port access city discussed in Section 4.2.1.

The Executive Officer (XO) or First Mate will serve as the Chief Operating Officer for the Earth Port. The four department heads will report directly to him/her: 2nd Officer in charge of the Deck Department; 3rd Officer in charge of the Administration Department; the Chief Supply Officer and the Chief Engineer.

5.2.2 Deck Department

The deck department will include personnel who load and unload cargo from aircraft, barges, and shuttle craft. It will include security and fire protection personnel.

5.2.3 Administration Department

This department will include the Medical, Dental, Human Resources, Information Technology, Chaplain, Safety and Customer Relations Divisions.

5.2.4 Supply Department

This department will include Shipping/Receiving, Food Service (Mess), and Hotel Division personnel.

5.2.5 Engineering Department

Under the Chief Engineering Officer, the Engineering Department will be responsible for the operation and maintenance of the equipment, facilities and systems aboard the FOP described in Section 4.3.5 and the tether terminus platforms described in Section 4.4.1. As is typical on a naval vessel or commercial cargo ship, Engineering Department personnel work regular shifts maintaining equipment and may also stand assigned watches in the Engineering Control Center. During daily climber operations on the tether platforms, a machinist and electrician will be assigned to each platform.

In the field of waste management, including recycling of solid wastes and the treatment/reuse of liquid waste, there will be specialized technicians trained to manage the on-board systems.

5.3 Climber and tether operating staff

The management and staff that carry out the basic mission of the Space Elevator enterprise at the Earth Port location are directed by HQ/POC personnel and supported by the crew of the OCC discussed in Section 4.3.4. The senior member, or Site Manager/Commander, will interface with the command staff of the FOP through its Captain and Executive Officer. Due to the highly technical and specialized skills required, the climber/tether operating staff may serve on a different rotational schedule or deployment than the crew of the FOP. This would be the equivalent of air squadron or wing operating from a naval aircraft carrier.

Coordination and cooperation between the climber/tether staff and the FOP crew will be of paramount importance in the daily activities of each. (See Section 6: "A Day in the Life.") As an example, FOP boatswains and/or helicopter pilots will transport the climber/tether staff to and from the terminus platforms. Deck hands from the Cargo Division will load the payloads aboard the service vessels and then offload at the tether terminus platform "just in time" for daily operations to begin. This coordination effort is reflected in the proposed organization chart shown in Figure 5-2.

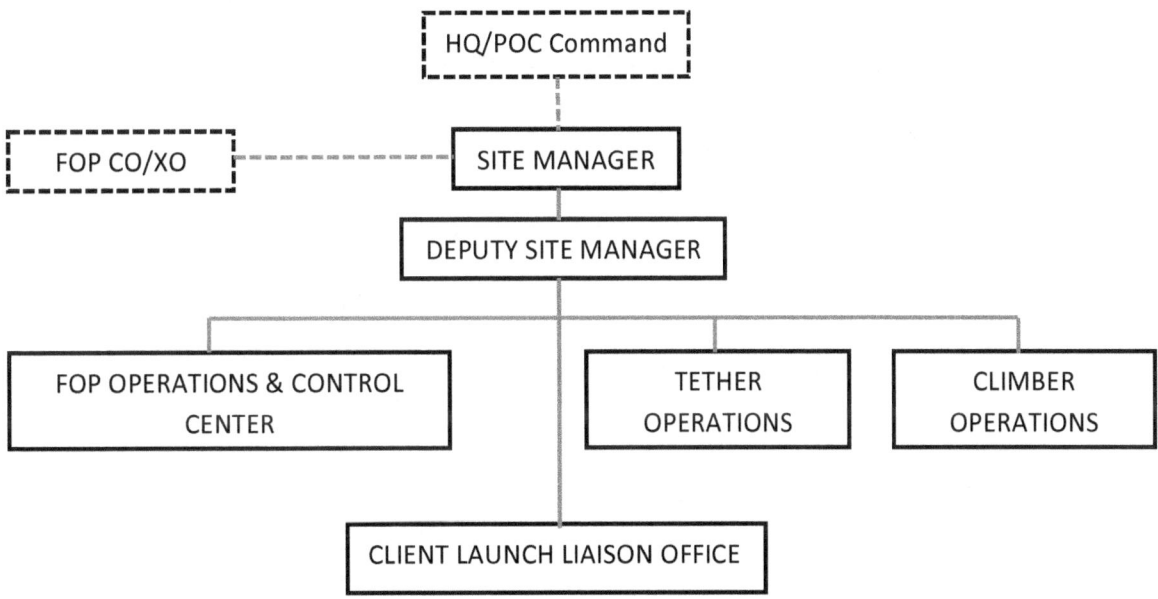

Figure 5-2 Climber/tether staff organization chart

5.3.1 Tether/Climber operations and control staff

5.3.1.1 Tether operations and control staff
Tether operations focus on correct mating and de-mating of climbers as well as monitoring the early part of a climb and the late part of a recovery. They also execute the reel-in and reel-out activities as directed by HQ/POC

5.3.1.2 Climber operations and control staff
The primary focus of climber operations is on mating and de-mating of climbers to the tether. Operators also monitor the early part of a climb and the late part of a recovery. Operators must additionally perform mating and de-mating of payloads with the climber. This may occur before or after mating with the tether. Climber operations would also be responsible for the assembly and installation of the protective shrouds.

5.4 Customers and visitors

The FOP and its crew will provide facilities and support for clients, visiting dignitaries including HQ executives and staff, U.S. and other nation's naval personnel, crews of OGVs servicing the platform, etc.

Considering steady state operations of the climbers, there might be 100 or more client personnel on board at any given time observing "their" individual operations. This is similar to today's payload launches. Each client could have up to ten executives and staff members in their party.

As a commercial enterprise, the FOP Captain, officers and crew must be prepared to "wine, dine and entertain" clients on board. These activities will be an extension of the client relationships established at the HQ/POC. FOP staffing, particularly in the Supply Department's Food Service and Hotel Divisions, must reflect this necessity.

5.5 Estimate of Earth Port operating costs

Below are estimates of the costs for a single Earth Port taken from the Concept of Operations (ISEC Position Paper #2013-1).

			Annual Cost	
	Staff of	29		
Fuel for Generators (3)	20 gallons per hour at $4 per gallon 500kw gen 3/8 loaded		$103,680	
Food & Water	Gallons per person per day	40	$423,586	
Supplies	Food @ $100 per person day	2900	$1,058,500	
Floating Platform	Assume large enough to provide living and office space for all operations and client/customer people. Storage for a climber or two and a satellite or two. 4 weeks of supplies and 2 climbers and 2 satellites. 4 weeks of fuel.			
Platform lease	5000 dollars per sq ft per month	100	$6,000,000	Saw them for sale in the $30-40 million range. Multiple LOPs increase cost linearly
Anually			$91,029,187	
OGV Operating Costs				
	$30,000 per day			
	10 typical number of days per trip			
	26 estimated trips per year		$7,800,000	
	Total FOP		$98,829,187	Multiple FOPs increase cost linearly

6. A Day in the Life (of a mechanic onboard the Earth Port)

Andy's alarm wakes him at 05:30 and he rolls out of his bunk. He hits the "head" to take care of his general daily grooming and then puts on his work coveralls. Before leaving, he hastily makes up his bunk and puts a few things away. He doesn't expect a stateroom inspection today, but he knows that a surprise inspection can be done at any time to check for fire and health code violations. He does a mental checklist to make sure he has everything he needs before starting work: ID badge? Check. Flashlight in case of a power outage? Check. Maintenance record tablet fully charged and secured in its holster? Check. Baseball cap? Check. He locks his room behind him and walks down to the chow hall for breakfast.

Almost everyone assigned to the morning crew shift is already there for breakfast, getting fueled up before their normal shift at 07:00 on the Operations Platform of the Space Elevator. Breakfast is pretty much the same each day; the hot line serves bacon, sausages, turkey sausages, scrambled eggs, hash browns, oatmeal, and cream of wheat, and the cold line carries items like pastries and fruit with an array of cereal boxes. Beverages are on a separate line featuring juice, milk, chocolate milk, and the never-ending supply of coffee. Andy goes through the hot line for eggs and sausage then grabs a coffee to take to a table.

While eating, he checks the monitor on the wall to see what the weather is like outside. Atmospheric conditions are constantly on the ticker that is running at the bottom of the screen. Current temperatures show 74 degrees with a high of 81. The main screen shows a communications crewmember who is acting as an activity coordinator, telling the rest of his shipmates what recreational activities are scheduled for that day. Andy considers seeing the movie being shown that night in the Recreation Center lounge.

After eating, he takes his tray to the bussing station to deposit leftover food and napkins in the green bin and his dishware on the conveyor belt to the scullery. At the end of the day, the contents of the green bin will go to the composting worms that turn it all into compost that is used to nourish the "rooftop gardens" that grow fresh vegetables, fruit, and herbs for the mess hall cooks to harvest. There is enough compost leftover that some of the crewmembers were allowed to start a separate garden to grow flowers, provided that it is a spare time activity that doesn't interfere with work or watch schedules.

Andy arrives at the Structures Shop and swipes his crew ID badge over the card reader by the door. He does this so that his paycheck will reflect what time he starts working, but it also functions as a method of taking muster—making sure no one is overboard. Of course, visual head-checks are done constantly to make sure someone isn't scanning another person's badge, but this is the primary form of tracking crewmembers on the Earth Port. The Structures Shop consists of the sheet metal shop, hydraulics shop, machine shop, weld shop, and the Non-Destructive Examiner (NDE) lab.

Andy and his crewmembers begin normal shift procedures by taking tool inventory, reading the pass down log left by the previous night shift to explain what was accomplished during their shift and what needs to be done next, and reading the POD (Plan of the Day.) The POD is a compilation of the Maintenance Schedule, Flight Schedule, Tether Schedule, and other important announcements concerning the crew. The department supervisor then goes to the maintenance meeting along with all of the other supervisor-level personnel to meet with the head of operations and receive his instructions for the day.

While waiting for him to return, each member of the Structures Shop logs into their respective maintenance tablets. They select open maintenance tickets that are within their scope of training. They are marked as "in work" so that the maintenance server can track the amount of man-hours used for each task and so the head of operations can see what is being worked on at any given moment. Each shop member then sets to their task: the machinist starts turning a replacement part on the lathe, the welder takes a part that needs repair behind a screen under an exhaust hood to add a new bead, a metalsmith follows a schematic to lay out a pattern on a large sheet of metal to fabricate a needed part, and the hydraulics guy starts sifting through the tubing to find one with the correct specifications to replace one that is leaking down in the power plant spaces. Andy inspects a basket of bolts for cracks in the Magnaflux machine. This is a Level 1 task, so he is able to do this one on his own. He can only do Level 2 work under the direct supervision of either the Level 2 or the Level 3 NDE, but he is working on his qualifications to become a Level 2 inspector, so he will be able to do those on his own, soon. Each person in the shop spends time learning how to assist others in their tasks for cross-training in case someone is sick or injured. The only jobs that require certification in his shop are NDE and welding, but they can still be done with direct supervision.

When the department supervisor comes back, he tells Andy to set up the vault for doing X-ray as soon at the bolts are done. The X-ray vault is merely a supply room converted into a vault by adding 3/8" thick lead sheets to the inside walls, floor, and ceiling to shield personnel from the radiation during a test. The department supervisor is also the Level 3 technician, so he will run the equipment while supervising Andy's on-the-job training. Andy is glad that the parts that need inspecting are small, otherwise they would have to arrange to use the helicopter pad when there is an opening in the launch schedule and cordon it off to everyone else onboard the ship to minimize exposure to the rest of the crew. That usually means between 03:00 and 04:00!

They complete their inspections a little after 08:30 and head back to the shop. There, Andy logs into the computer to enter the results of the inspections and attaches each of the digital radiograph files to its respective maintenance request so that the parts can be returned to service. The supervisor sends Andy to train with the hydraulics specialist to learn how to cut and shape tubing and attach fittings to them. Andy already has his Airframes and Power Plant license, but doesn't have much experience with tubing, so he will assist until he's ready to work on his own. After the parts are made, Andy checks back in with the supervisor and is assigned as the fire watch for the welder while she works on a stack of parts. Andy continues to work on jobs like this until lunchtime.

At the chow hall, Andy is not enthusiastic about the beef and gravy over buttered noodles (again!) offered on hot line #1 and heads over to line #2, which always has burgers, hot dogs, and pizza. He grabs a couple of slices of pepperoni pizza and a soda from the beverage line and sits with some crewmembers that he has made friends with. After lunch, he has a few more minutes before he has to be back at the shop, so he heads down to the ship's store to pick up some snacks. He asks if the cribbage board he ordered will be on the next supply ship, which arrives in another week. The supply clerk in charge of the store checks her terminal and verifies that it will be in the next shipment. Andy looks forward to challenging some of his crewmates who also claim to be pretty good at Cribbage to a few games in the Recreation Center.

When Andy gets back from lunch, he is assigned to do more NDE inspections. This time, he uses an ultrasonic machine to test for de-laminations in some composite panels. After the supervisor checks his results, Andy enters the results of the inspection into the maintenance tablet. Andy spends the rest of his day in the shop learning new skills and honing old ones.

At about 14:30, crewmembers from the night shift filter in to the shop to begin pass down procedures. Andy and the Level 2 NDE technician assigned to the night shift decide to do the tool inventory, so they walk around to each tool station to verify that all parts are either present or accounted for. After they both register their electronic signatures on their maintenance tablets to verify that the task was completed, the Level 2 tech reads the pass down and POD. Andy sweeps the floor and waits for the supervisor to signal the end of their regular eight hour shift just before 15:00. Andy signs out with his badge and heads to his stateroom.

There, he takes a shower and puts on fresh clothes to rid himself of soot, hydraulic fluid, and metal shavings before heading to the chow hall for dinner. The guys at his table ask him if he is going to watch that movie in the Rec Center lounge that was advertised on the activity video. He agrees to meet the rest of them there after working out. Andy spends about an hour in the gym, then heads back to his stateroom for a short, refreshing shower and another change of clothes. He heads to the Rec Center for the movie that starts at 18:30. After it ends, Andy checks the time, and, since it is only 20:30, he knows he will have time to get to the weather deck of the Operations Platform to watch the launch of the Space Elevator scheduled to go up at 21:00. He saunters out to join the small crowd gathering on the part of the deck that faces Tether Platform A.

Andy has seen many of these launches in the last few months of being on-station, but it never seems to get old. Anyway, it's an excuse to get outside for some fresh air. After a few minutes, they start to hear some activity on the radio that someone brought out to listen in on operations. Within another ten minutes, they see the elevator lift off for space. Andy is able to watch it for a few more minutes before it dwindles to the size of pea and the tether is no longer visible. If he had binoculars, he'd be able to watch it for a while longer, but the novelty of that wore off a long time ago. With a smile, he remembers his first time watching a Space Elevator launch. He stood staring at the sky with his mouth agape long after he could no longer see it, hoping to catch a reflection of light off of a panel. Soon, the next high-speed ferry will arrive with the next set of new personnel and the seasoned crewmembers will smile as they see the launch through the eyes of a "newbie" while listening to them "ooh" and "aah" as they stare at this technological wonder in amazement for their first time. Andy follows a small crowd through the watertight door and down the passageway until he gets to his stateroom to retire for the evening.

Zero five-thirty comes early, so Andy gets to sleep at a decent hour, letting the platform's gentle rocking lull him to sleep, resting up for another busy day in support of the Space Elevator Earth Port.

7. Roadmap for Earth Port Development

ISEC study report 2014-1 showed that the needed technology is feasible and has a valid engineering application. At or below the segment level, tests, experiments, and/or demonstrations are required. From the Roadmap Report the following is suggested:

The Earth Port Segment's path is seen as resolving five primary (Earth Port) functions:

- serves as Space Elevator tether terminus including reel in/out (tension, wind, current, debris avoidance), and position management;
- serves as a port for receiving and sending ocean-going vessels. The OGV's that come and go from the Earth Port are moving climbers, payloads, supplies and personnel;
- serves as facility for attaching and detaching payloads to and from tether climbers and attaching and detaching climbers to and from the tether;
- provides tether climber power for the 40 km above the FOP;
- provides food and accommodation of crew members as well as power, desalinization, waste management and other such support.

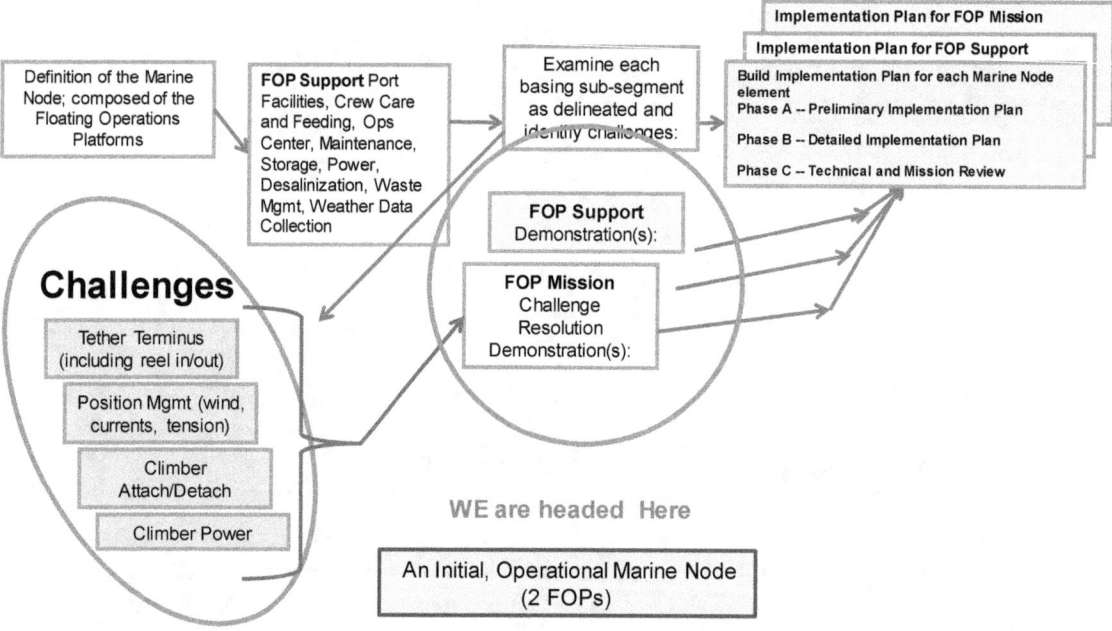

Figure 7-1 Earth Port Roadmap

As indicated by the roadmap (see full ISEC 2014 study report) and to demonstrate four of the five primary roles, an ordered taxonomy of sequenced tests will be conducted. The care and victualing of crew members is not considered a "challenge". In support of this test activity a wide-ranging survey must be conducted related to Earth Port site location, ocean currents, surface wind speeds, water temperature and salinity variations, sea floor geology and weather information. Ultimately, the validity of the Earth Port will be shown in four culminating demonstrations.

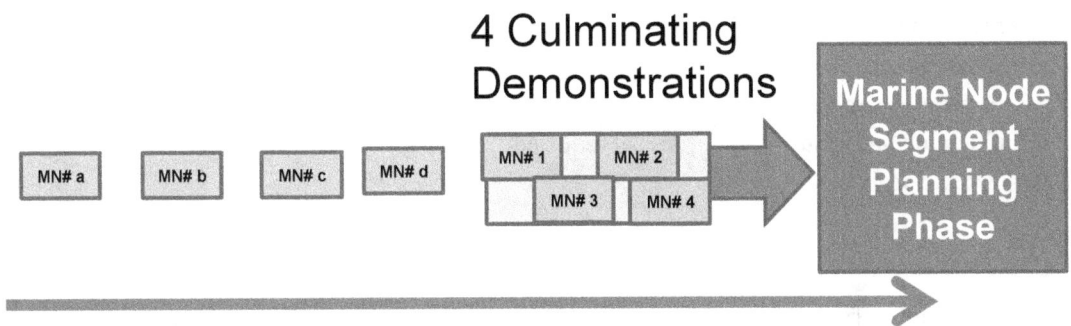

Figure 7-2 Marine Node Pathway

Term	Demonstration	Dependency
MN-1 Far	Tether Terminus (includes reel in/out)	Needs Tether
MN-2 Mid	FOP Position Management	Needs Tension spec
MN-3 Far	Climber Attach/Detach	Needs Tether
MN-4 Far	Climber Power	Needs Climber

Table 7-1   Marine Node culminating demonstrations

Now that the Earth Port has taken form in this report, the development team will need to consider what other tests, experiments or demonstrations are needed to adjudicate the Earth Port segment and (especially) its trade-off considerations and overall integration as a singular element within the Space Element Architecture. Specific Earth Port tests, experiments, and demonstrations must be developed to ensure that they meet the objectives of the roadmap flow. Some are listed below.

1. Earth Port facility testing; essentially a flow of materials handling tests
2. Mobile container crane, payload handling
3. WAVEWATCH data versus situational awareness
4. Multiple and varying reel-in and reel-out tests
5. Floating wave attenuator in various seas
6. Debris management
7. Tether terminus operations
8. Power allocation and distribution at levels required
9. And others as needed

The demonstration line portrayed in the pathway shown above is essentially a test campaign conducted at each step with entry and exit conditions cited as part of the campaign execution. Note that the test campaign will focus on the Demonstrations associated with the challenges shown in the roadmap in Figure 7-1. "Demonstrations" should include a range of tests, inspections, analyses, simulations, and more. Some are likely to be a sequence of test events; a taxonomy of tests. The Term for each is expressed as: Near (2014-2020), Mid (2020-2028), Far (2028-2035). A suggested list of culminating demonstrations was shown in the 2014 study.

The Earth Port will also be the base for any number of operational tests during the transition toward IOC. In that sense, the Earth Port holds a unique position as the center of activity for test and demonstration practice and training. Given the idea of a minimum IOC system to be built first, which then grows into something robust, the role of the Earth Port is still to be defined. In addition to being the center of operational and transitional testing, it is likely that the Earth Port will see training on its list of functional responsibilities.

8. Conclusions and Recommendations

The authors conclude that:

1. **Design and construction of the Earth Port will be an extension of today's practices**. As the Earth Port consists of many components, there will be a systems engineering design aspect that will ensure consistency between the Floating Operations Platform (FOP) and logistics flow of support and mainline mission equipment. The Earth Port has a tremendous amount of historical experience to leverage. The design of floating platforms goes back thousands of years while the ability to operate from them is mature. In addition, there are many parallel developments of operations centers that can help design the FOP. The roadmap, accomplished for the Earth Port in the 2014 ISEC study [Space Elevator Architecture and Roadmaps], shows that development can be accomplished in a linear manner with current technologies.

2. **Operations at the Earth Port will leverage over a hundred years of experience**. This expertise and knowledge will come from off-shore drilling operations and naval operations around the world. The FOP will likely be a modified drilling platform to support the combined need of securing space elevator tethers and enabling the processing of payloads through its transportation hub. Each of the major components of the Earth Port will require support from experienced personnel who have worked inside off-shore drilling platforms and transportation hubs.

3. **Operation and maintenance costs appear to be reasonable compared to similarly sized projects**. By using parallel constructs, the estimate of staffing and platform sizes has enabled this team to estimate the costs. The key here is to ensure that the operations are routine and established early in the development of our space elevator concept. Most of the processes must be straightforward to ensure that the routine movement of payloads and tether climbers is enabled in a rapid manner supporting needs of customers. There will be a business center at the Headquarters that is responsible for ensuring the costs are estimated correctly and then adhered to.

Recommendation: As each of the International Space Elevator Consortium's year-long study finishes, it is obvious that we have just initiated the discussions on the topic. This team recommends that efforts continue to refine the ideas in this report to help reduce risk for future development.

Appendix A   International Space Elevator Consortium

Who We Are

The International Space Elevator Consortium (ISEC) is composed of individuals and organizations from around the world who share a vision of humanity in space.

Our Vision

A world with inexpensive, safe, routine, and efficient access to space for the benefit of all mankind.

Our Mission

The ISEC promotes the development, construction and operation of a space elevator infrastructure as a revolutionary and efficient way to space for all humanity.

What We Do

- Provide technical leadership promoting development, construction, and operation of space elevator infrastructures.
- Become the "go to" organization for all things space elevator.
- Energize and stimulate the public and the space community to support a space elevator for low cost access to space.
- Stimulate science, technology, engineering, and mathematics (STEM) educational activities while supporting educational gatherings, meetings, workshops, classes, and other similar events to carry out this mission.

A Brief History of ISEC

The idea for an organization like ISEC had been discussed for years but it wasn't until the Space Elevator Conference in Redmond, Washington, in July of 2008, that things became serious. Interest and enthusiasm for a space elevator had reached an all-time peak and with Space Elevator conferences upcoming in both Europe and Japan, it was felt that this was the time to formalize an international organization. An initial set of directors and officers were elected and they immediately began the difficult task of unifying the disparate efforts of space elevator supporters worldwide.

ISEC's first Strategic Plan was adopted in January of 2010 and it is now the driving force behind ISEC's efforts. This Strategic Plan calls for adopting a yearly theme to focus ISEC activities. (For 2010, the theme was "Space Elevator Survivability - Space Debris Mitigation.") In 2010, ISEC also announced the first annual Artsutanov and Pearson prizes to be awarded for "exceptional papers that advance our understanding of the Space Elevator." Because of our common goals and hopes for the future of mankind off-planet, ISEC became an affiliate of the National Space Society in August of 2013.

Our Approach

ISEC's activities are pushing the concept of space elevators forward. These cross all disciplines and encourage people from around the world to participate. The following activities are being accomplished in parallel:

- CLIMB – This annual peer reviewed journal invites and evaluates papers and presents them in an annual publication with the purpose of explaining technical advances to the public. The first issue of CLIMB was dedicated to Mr. Yuri Artsutanov (a co-inventor of the space elevator concept) and the second issue was dedicated to Mr. Jerome Pearson (another co--inventor). CLIMB is scheduled for publication each July.

- Yearly conference – International space elevator conferences were initiated by Dr. Brad Edwards in the Seattle area in 2002. Follow-on conferences were in Santa Fe (2003), Washington DC (2004), Albuquerque (2005/6 –smaller sessions), and Seattle (2008 to the present). Each of these conferences had multiple discussions across the whole arena of space elevators with remarkable concepts and presentations. Recent conferences have been sponsored by Microsoft, the Seattle Museum of Flight, the Space Elevator Blog, the Leeward Space Foundation, and ISEC.

- Year-long technical studies – ISEC sponsors research into a focused topic each year to ensure progress in a discipline within the space elevator project. The first such study was conducted in 2010 to evaluate the threat of space debris. The second study, and resulting report, focused on space elevator operations. The 2013 study focused upon tether climber designs. The 2014 topic was Space Elevator Architectures and Roadmaps. There were two topics chosen for 2015: Marine Node Design Considerations and Status of Tensile Strength materials development. The products from these studies are reports that are published to document progress in the development of space elevators.

- International cooperation – ISEC supports many activities around the globe to ensure that space elevators keep progressing towards a developmental program. International activities include coordinating with the two other major societies focusing on space elevators: the Japanese Space Elevator Association and EuroSpaceward. In addition, ISEC supports symposia and presentations at the International Academy of Astronautics and the International Astronautical Federation Congress each year.

- Competitions – ISEC has a history of actively supporting competitions that push technologies in the area of space elevators. The initial activities were centered on NASA's Centennial Challenges called "Elevator: 2010." Inside this were two specific challenges: Tether Challenge and Beam Power Challenge. The highlight came when Laser Motive won $900,000 in 2009, as they reached one kilometer in altitude racing other teams up a tether suspended from a helicopter. There were also multiple competitions where different strengths of materials were tested going for a NASA prize – with no winners. In addition, ISEC supports the educational efforts of various organizations, such as the LEGO space elevator climb competition at our Seattle conference. Competitions have also been conducted in both Japan and Europe.

- Publications – ISEC publishes a monthly e-Newsletter, its yearly study reports and an annual technical journal [CLIMB] to help spread information about space elevators.   In addition, there is a magazine filled with space elevator literature called Via Ad Astra.

- Reference material – ISEC is building a Space Elevator Library, including a reference database of Space Elevator related papers and publications.

- Outreach – People need to be made aware of the idea of a space elevator. Our outreach activity is responsible for providing the blueprint to reach societal, governmental, educational, and media institutions and expose them to the benefits of space elevators. ISEC members are readily available to speak at conferences and other public events in support of the space elevator. In addition to our monthly e-Newsletter, we are also on Facebook, Linked In, and Twitter.

- Legal – The space elevator is going to break new legal ground. Existing space treaties may need to be amended. New treaties may be needed. International cooperation must be sought. Insurability will be a requirement. Legal activities encompass the legal environment of a space elevator - international maritime, air, and space law. Also, there will be interest within intellectual property, liability, and commerce law. Starting work on the legal foundation well in advance will result in a more rational product.

- History Committee – ISEC supports a small group of volunteers to document the history of space elevators. The committee's purpose is to provide insight into the progress being achieved currently and over the last century.

  Research Committee – ISEC is gathering the insight of researchers from around the world with respect to the future of space elevators. As scientific papers, reports and books are published, the research committee is pulling together this collective progress to assist academia and industry to progress towards an operational space elevator infrastructure. For more, visit http://isec.org/index.php/about-isec/isec-research-committee

ISEC is a traditional not-for-profit 501 (c) (3) organization with a board of directors and four officers: President, Vice President, Treasurer, and Secretary. In addition, ISEC is closely associated with the conference preparation team and other volunteer members.

Address: ISEC, 16991 McGill Road, Saratoga, CA 95070 / http://www.isec.org

Appendix B Acronyms and Lexicon

CNT	Carbon Nano Tube
CO	Commanding Officer
FOP	Floating Operations Platform
GEO	Geosynchronous Earth Orbit
HQ/POC	Headquarters Primary Operations Center
IAA	International Academy of Astronautics
ISEC	International Space Elevator Consortium
kg	kilogram
LOA	Length Overall
MT	metric ton
MW	megawatt
NASA	National Aeronautics and Space Administration
NCEP	National Centers for Environmental Prediction
NOAA	National Oceanographic and Atmospheric Administration
OCC	Operations Control Center
OGV	Ocean Going Vehicle
STOL	Short Takeoff & Landing
VTOL	Vertical Takeoff & Landing
XO	Executive Officer

IAA study group #3-24 met in Seattle in August of 2015. The team agreed to use, as much as possible, consistent terminology for this report. These terms are listed in the table below and shown in Figure B-1. A table of general, suggested terminology with explanations follows the figure.

Apex Anchor Node	LEO Gate	Earth Port
Mars Gate	Lunar Gravity Center	- Earth Terminus
Moon Gate	Mars Gravity Center	- Floating Operations Platform
GEO Node	Tether Climbers	Headquarters and Primary Operations Center

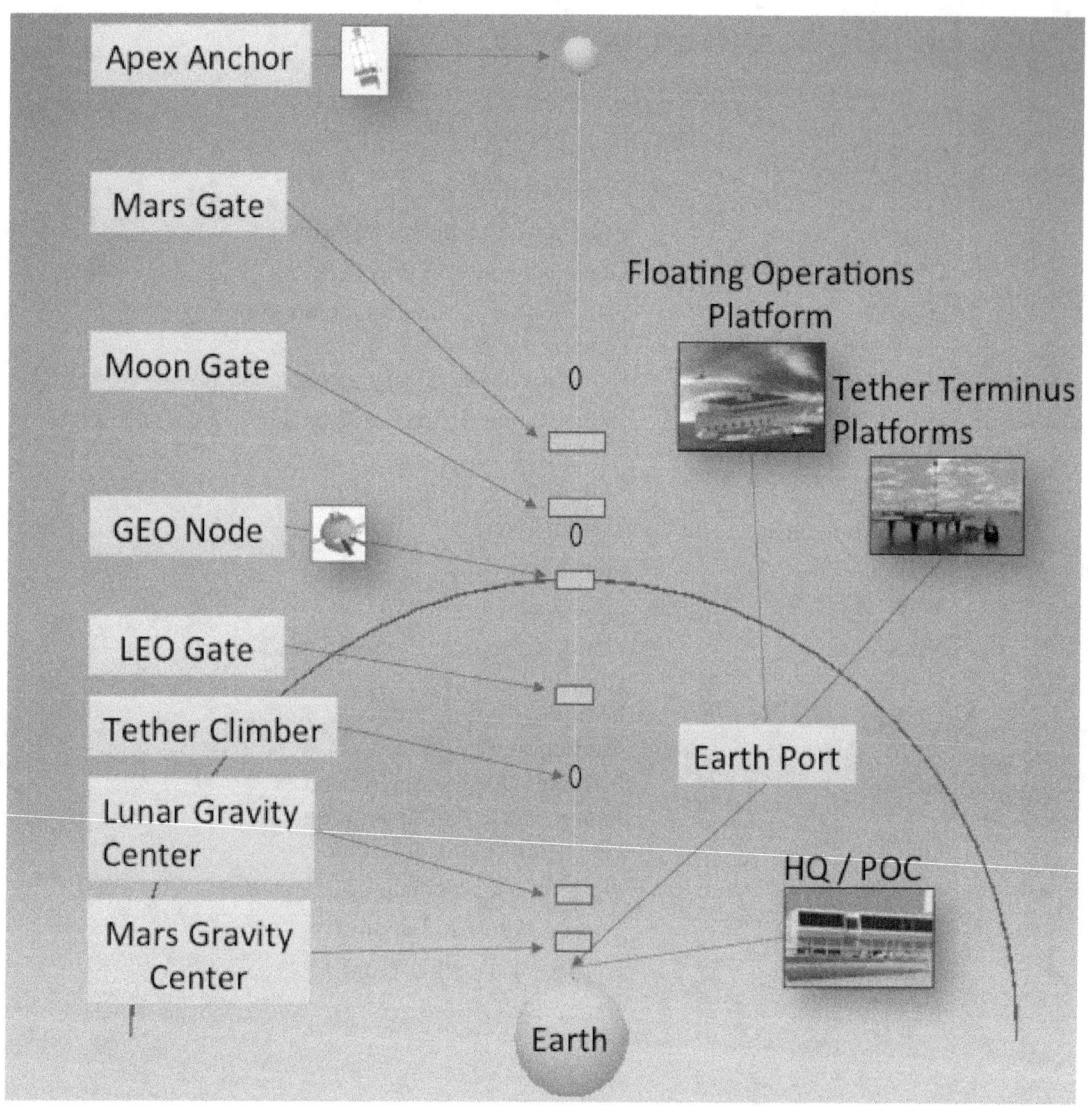

Figure B-1 Space Elevator System

Table of Suggested Terminology

Terminology	Explanation
Apex Anchor	The upper end at roughly 100,000 km altitude
Carbon Nano Tube	High Tensile Strength material under development
Climbers [Tether Climbers]	Vehicle able to climb or lower itself on the tether
Deployment	Releasing the tether from the GEO construction up and or down during the initial phase of construction
Earth Anchor	Earth Terminus for space elevator
Earth Port	Consists of Earth Anchor [terminus] and Floating Operations Platform
Final Operational Capability	Design for full capability of the space elevator
Floating Operations Platform	The Ops Center for the activities at the Marine Node or Earth Terminus [Earth Port]
GEO Node	Geosynchronous Earth Orbit (GEO) Release Point – roughly 36,000 km
Headquarters and Primary Operations Center [HQ/POC]	Location for the Operations and Business Centers – probably other than on Earth Port
Initial Operational Capability	A term to describe the time when the space elevator is prepared to operate for commercial profit - robotically
International Space Elevator Consortium	Association whose vision is to: A world with inexpensive, safe, routine, and efficient access to space for the benefit of all mankind.
Japanese Space Elevator Association	JSEA handles all the space elevator activities for universities and STEM activities. Also handles the global aspects of space elevators.
Length Overall	Full length of the space elevator, est. from 96,000 to 100,000 km
LEO Gate	Elliptical release point for LEO – roughly 24,000 km
Lunar Gravity Center	Point on Tether with Lunar gravity similarity – 8,900 km
Earth Port	Earth Terminus for space elevator
Mars Gate	Release Point to Mars – roughly 57,000 km
Mars Gravity Center	Point on Tether with Mars gravity similarity – 3,900 km
Moon Gate	Release Point towards Moon – roughly 47,000 km
Ocean Going Vehicle (OGV)	Vehicle able to travel over the open ocean
Primary Operations Center	Center of all activities for the space elevator. Could be distributed or centralized.
Floating Operations Platform (FOP)	Operations center at the Earth Port
Seed Ribbon	The initial tether lowered from GEO altitude which would then be built up to become the space elevator tether
Tether	100,000 km long ribbon of space elevator
Tether Climber	Vehicle able to climb or lower itself on the tether

Appendix C   Earth Port Air and Sea Vehicles

C.1 Ocean-going tug/barge

The primary OGV for delivering customer payloads and supplies to the Earth Port from the access cities port facilities will be regularly scheduled ocean-going barges powered by large ocean-going tugs. Although their speed is slow (10 knots), their range is large and the on-deck cargo capacity will easily meet the needs of space elevator operations. It is anticipated that the round trip from access port to the Earth Port's FOP will be 9-10 days. Ocean-going tugs are typically powered by conventional marine diesel fuel and would not need refueling at the FOP. Two or three chartered tug/barge combinations on regular staggered schedules should be sufficient to meet the needs of Earth Port operations. A typical ocean-going tug with barge is shown in Figure C-1.

Typical Ocean-going Barge

400 feet length overall (LOA) by 105 foot beam by 25 foot draft

Typical Ocean-going Tug

140 feet LOA by 40 foot breadth by 15 foot draft

Figure C-1   Ocean-going tug and barge

As an alternative to the tug/barge combination, older generation small general cargo ships capable of handling containerized cargo may be utilized as the primary cargo carrier for the Earth Port.

C.2 High-speed ferries and long-range seaplanes

The location of the Earth Port relative to major coastal cities puts it beyond the range of most large capacity helicopters (600-1000 miles). Therefore, the majority of people working aboard the Earth Port platforms, as well as clients and visitors, will access the Earth Port using either a high-speed ferry, VTOL or long-range seaplane. These modes of travel may also serve to transport loads which are smaller than container-sized and of high value and/or otherwise sensitive cargo. Consumable supplies and operating equipment for the FOP may also be carried on ferries. Cargo capacity may be up to 500 MT.

A typical high-speed ferry has twin aluminum hulls powered by gas turbine engines. Cruising speeds average 50 knots. Stabilizing fins produce a relatively smooth ride for its passengers (see Figure C-2). Comfortable seating, food service, entertainment, satellite communications and overnight accommodations are provided. The trip length from the access city to the Earth Port will depend upon final locations, but there will likely be an attempt to keep it to a maximum of one day. The ferries will be refueled at the FOP for the return trip.

High-speed Ferry

170 feet LOA by 38 foot beam by 10 foot draft. Air draft (height above MSL) 70 feet.

Ocean going (super) yachts may also call upon the FOP as clients and visitors may wish to utilize their own vessels.

Figure C-2   High-speed ferry

As an alternative to high-speed ferries and other watercraft to transfer personnel to/from the Earth Port, long-range amphibious planes may be used. As a current example, Shinmaywa Industries of Japan has developed the US-2 Short Takeoff and Landing (STOL) amphibious plane (see Figure C-3). Its cruising speed is 260 nautical miles per hour and its range is 2,400 nautical miles. It can carry up to 20 passengers and 18 MT of cargo. With this speed and range, the trip from the access city to the Earth Port will take between four and five hours. The seaplane has an overall length of 110 feet and a wingspan of 109 feet. Earth Port will contain refueling and maintenance facilities for aircraft.

Figure C-3   Long-range amphibious airplane

As described in Section 4.3, the cargo will be moved, after appropriate processing, from the FOP to one of the tether terminus platforms for placement into the climber. Assuming loads up to 14 metric tons, an offshore platform service vessel may be used to perform this function (see Figure C-4). These vessels will probably be the "workhorses" of the Earth Port, performing multiple operational functions beyond moving cargo and personnel between the floating platforms.

Offshore Service Vessels

140 feet LOA by 29 foot beam by 13 foot draft. Air draft is 93 feet

Figure C-4   Offshore service vessel

These vessels are powered by conventional marine diesel engines and can carry deck loads of up to 200 MT. Its working speed within the Earth Port exclusion zone will be 10-12 knots and its total horsepower is around 7,200 bhp. It is outfitted with bow and stern thrusters for station-keeping. This workboat may also be outfitted with firefighting pumps and nozzles. For operational reliability and other reasons, the Earth Port fleet should have several of these vessels.

C.4 Harbor tugs and small craft

As discussed in Section 4.4, the tether terminus platforms may need to move off station or reorient on station with relatively short notice from the FOP operations center. To assist in these movements and to assure necessary positioning for the free-floating, semi-submersible tether terminus platforms, it may be advisable to include two harbor tug-type vessels in the Earth Port fleet. (Figure C-5). To augment the tether terminus platforms' on-board thrusters, the tugboats would be secured by lines to the platforms.

Harbor Tug

80 foot LOA by 34 foot beam by 15 foot draft. Air draft is 60 feet. Power is provided by marine diesel engines up to 10,000 bhp. These boats would be outfitted with fire-fighting pumps and nozzles and would serve as the principal fireboats for the Earth Port.

Figure C-5 Harbor tugboat

The ultimate dimensions of the Earth Port's Exclusionary Zone (see dotted circle on Figure 4-2.) will be determined by several factors, including the dynamics of the tether terminus platform operations. Within this "circle," surface, airspace and underwater security will be paramount. Complementing the long-range and near-field radar and sonar systems operating continuously on the FOP, direct surface-level security will be provided by high-speed surface patrol craft (Figure C-6).

Figure C-6   High-speed patrol craft

A typical harbor patrol craft has dimensions of 40 feet LOA by 13 foot beam by 2 foot draft. These boats have aluminum hulls driven by diesel or gas marine engines and jet drives. Top speed is 40 knots with a range of 250 nautical miles. They can be outfitted with appropriate communications equipment, weaponry and fire-fighting capabilities. They may also be used to ferry crews back and forth between the FOP and the tether terminus platforms to supplement the helicopter service.

The Earth Port will also have a number of Zodiac™ type outboard utility boats within its fleet. These craft (an example is shown in Figure C-7) can be used for everything from deploying and tending the floating breakwaters when necessary to recreational uses for FOP staff and visitors such as fishing and waterskiing.

Figure C-7 Outboard utility boat

C.5 Short-range helicopters and amphibious patrol planes

Personnel involved in the primary operation of lifting and retrieving payloads from the climbers will routinely be transported from the FOP to the tether terminus platforms by short-range helicopters. Obviously it is premature to specify the dimensions and capabilities of these aircraft. However, they should be able to handle up to six passengers.

For security purposes it may also necessary to deploy short-range amphibious patrol planes within and beyond the boundaries of the Earth Port Exclusionary Zone. Drone-type aircraft may also be used for perimeter security. See Figure C-8.

Figure C-8 Earth Port aircraft

Appendix D High Stage One

High Stage One of the space elevator supports a platform at 40 km altitude where payloads are transferred from a funicular railcar to a tether climber for onward travel up the tether towards the GEO node (see **Figure**). It deals with the sometimes large forces generated by winds and ice, transferring these forces to the Earth's surface. This relieves the tether of the fluctuating tensions caused by Earth's turbulent atmosphere. Otherwise, these forces would at times more than triple the load on the tether, if the tether reaches down to the surface.

High Stage One is an adaptation of the Launch Loop originally proposed by Keith Lofstrom as a direct-to-orbit launch mechanism. High Stage One is roughly 1/12 the size of the Launch Loop but uses the same concept of holding up the structure by changing the direction of the momentum vectors of rotors traveling inside evacuated tubes using magnetic levitation to minimize friction.

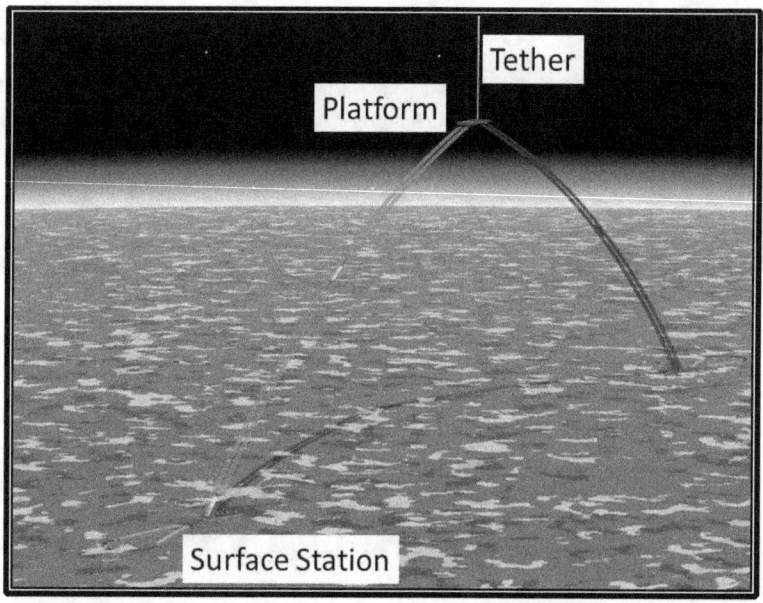

Figure D-1 High Stage One

Another benefit of High Stage One is that it is easy to move the tether to avoid space debris, because the base of the tether is in very thin air. It is possible to swing a 5-ton weight from the platform, causing a lateral wave to be propagated up the tether in the required direction. By contrast, moving it at the surface involves getting a ship underway, and that would sometimes have to be against opposing storms.

Each of the two surface stations has to sustain a load of 27,000 metric tons to support everything above it. This is not particularly large for ocean-going vessels, but it is a different requirement from the Marine Node where the tether reaches to the surface. Position-keeping is still required so as to maintain the overall stability. Depending on the depth and the state of the ocean bottom, the surface station may use anchors, active position-keeping, or a combination of the two. **Figure** shows more detail of a surface station. Most of the weight is experienced at the ramp, which will either be suspended from the ship or will have its own flotation supports. There is also a large horizontal force tending to pull the ambit away from the rest of the surface station. This could be sustained by anchors or by long underwater cables joining the two surface stations, which are 112 km apart.

One of the surface stations will house the Floating Operations Platform (FOP). High Stage One requires the FOP to manage the following two additional tasks beyond those needed for the Earth Port with the tether reaching the surface.

1. The transfer of payloads to tether climbers on the platform: the funicular can raise payloads at any time, but the tether climbers can only commence climbing at dawn, due to their need for solar power, or at the time dictated by the disposition of other tether climbers, so as to manage the overall load on the tether. Climbers will be loaded and prepared on the platform ready for departure. The platform at 40 km altitude is effectively a spacesuit environment. Most operations will be handled remotely from the FOP. Service personnel will be able to make visits for maintenance purposes.

2. Maintenance and health checks on High Stage One: Every component of the rotors traveling at high speed inside the evacuated tubes will be checked automatically, with the results being monitored at the FOP. The structure consists of five pairs of tubes. Routine maintenance will involve bringing down one pair of tubes for servicing while the remaining four pairs continue to operate and support the platform and funicular.

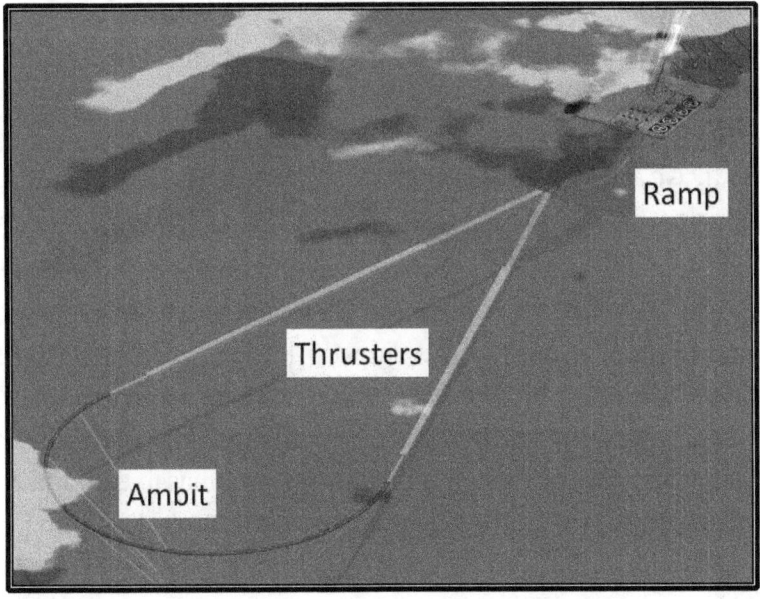

Figure D-2 Detail of surface station

Appendix E Mini Workshops Results

A mini-workshop on the space elevator marine node was held at the August 2015 ISEC conference in Seattle. The Earth Port is the interface between the space elevator and the Earth's surface where passengers and payload will be transferred from surface vessels or aircraft to tether climbers. The workshop focused on four areas of study: the functional requirements of the marine node, its physical plant and operations, its organization and staffing, and the culminating demonstrations required to prove the feasibility of its key aspects.

The workshop commenced with a talk by Vernon Hall summarizing the progress of the Marine Node Study begun in the Fall of 2014. The report includes discussions and conclusions concerning the need for a space elevator and what a typical day of space elevator operations would be like. The marine node as currently envisioned consists of three floating platforms: a Floating Operations Platform (FOP) and two tether terminals, all connected by a fleet of ocean-going vessels (OGV), helicopters and seaplanes. The first 40 km of each tether is also included as part of the marine node. The functional requirements include node location, methods of tether control and payload attachment, transport and service logistics, communications and security. The FOP is planned as new, purpose-built, construction, but early versions of it could rely on conversion of existing vessels or platforms. The tether terminal would likely be platforms very similar to offshore oil rigs or aircraft carriers. The marine node will require a mix of existing and yet-to-be-developed technology which will have to be prototyped and demonstrated.

The Earth Port report will be supplemented by results from the workshop discussions which examined the tether terminus platforms, the FOP and the culminating demonstrations.

The tether terminus is a complex interface between the mechanical aspects of the tether and the surface operations required to run and service it. Each terminus must provide communications, power and tether control up to an altitude of 40 km. How the tether is attached to the platform will affect the methods of controlling its motion, which would include pulling on the tether or reeling it in and out. The current plan sites the platforms on the equator to provide maximum payload capacity, but there are weather, tether control and debris avoidance advantages to siting them off the equator by as much as 8 degrees. Efficient methods of payload assembly and transfer from platform to climber must be developed. Regular maintenance of climbers, tethers and platforms must be planned; this entails a large, on-site store of spare parts as on-time delivery from the distant shore would be costly or unlikely. The platforms must be protected from their physical environment. Although the platforms are massive and thus very stable under most conditions, floating breakwaters would mitigate wave action. Lightning protection for both platform and tether will be required and emergencies such as a tether break, sinking platform or hostile action should be anticipated.

The FOP serves as the forward base of operations and seaborne port which will connect to the primary operations center (POC) and mainland cities. It will coordinate inbound and outbound traffic and distribute it to the tether terminus. A land-based FOP was discussed as having greater physical stability, lower construction costs and shorter supply lines, but these advantages must be balanced against safety, weather and national jurisdiction concerns. The FOP will be the nexus for local communications and security operations as well as cargo handling and storage. The staffing levels required to supply these functions are expected to be high, and difficult living conditions must be taken into account. Robotics and tele-operation could reduce this level significantly. The concerns of space elevator investors and the comfort of passengers must be taken into account early in the FOP design stage. On-site power generation will be necessary unless the FOP is located near a shore. A wide range of options was discussed, including custom-built nuclear plants, existing US Navy plants, harnessing wave action or ocean thermal energy, or tapping undersea oil or geothermal power.

Before being built or implemented, the feasibility of each key element of the space elevator must be demonstrated. Several such culminating demonstrations were discussed, including the tether terminus concept, tether position management, the climber attachment/detachment procedures and power generation. It must be shown that tether tension management is possible and that the relative position and tension of tether elements can be accurately monitored. Position management can be demonstrated using control and guidance systems which rely on physical registration markings on the tether. The dynamic effect of attaching and detaching climbers to the tether must be demonstrated and the attachment/detachment procedure must show that several different types of climber can be accommodated. Currently only payload and repair climbers are envisioned for standard operations, but others are likely. A proof of concept for power generation for the terminus and the first 40 km of tether must show that power can be efficiently produced and distributed, taking into account surge protection, solar disruptions, and emergency power loss scenarios.

Once these challenges are met the Earth Port will become the spaceport of the future. It will grow from a small base into a busy commercial hub and eventually into a floating city with a large range of services and a sizable population.

Appendix F Howland Island Alternative

Introduction

During the Marine Node workshop held at the 2015 Conference in Seattle, the concept of using land-based facilities for some or all of the Marine Node operating requirements was once again discussed (see Appendix E). It was recommended by Dr. Robert G. Williscroft, Ph.D., former U.S. Navy and NOAA officer and deep sea diver, that ISEC consider Howland Island and vicinity as the site for the Earth Port as an alternative to floating platforms in the Pacific Ocean on the equator. This appendix explores the Howland Island environs and briefly compares the differences in transportation, facilities and operations between such a location and that described in Section 4 above.

Howland Island description and location

Howland Island is an uninhabited coral island and is an unincorporated territory of the United States. It covers 450 acres (1.8 ha) and has 4 miles (6.4 km) of coastline. The terrain is low-lying and sandy with its highest point about 6 meters above mean sea level. It has no natural fresh water and is primarily a resting and nesting area for a variety of seabirds.

Figure E-1 Howland Island looking northerly

The island is currently managed by the U.S. Fish and Wildlife Service as part of the Pacific Remote Islands Marine National Monument. Water depths within the claimed 200 nautical mile Exclusive Economic Zone around the island are in the range of 6,000 feet. The island is about 1 kilometer wide (3,300 feet) with a long north-south axis of about 2.4 kilometers (7,870 feet.) It lies about 55 nautical miles north of the equator in the Pacific at 0° 48' 24"N latitude and 176° 36' 59" W longitude.

Transportation considerations

Section 4.2.1 of this report discusses the concept of an Earth Port access city and HQ/POC location that meets several criteria including air and sea distance to the Earth Port that should be within one day's travel time for passengers. The closest world city to Howland Island that meets all of the criteria would be Honolulu, HI. (21° 18' 37" N, 157° 52' 15"W.) The distance between Honolulu Harbor and Howland Island is 1,650 nautical miles.

The ocean-going vessel (OGV) fleet discussed in Section 4.2.2 describes an ocean-going tug/barge combination as the primary service vessel for cargo and payload to/from the Earth Port. This transportation method would still be suitable for the Howland Island alternative.

The use of high-speed ferries to transport personnel and smaller quantities of high-value cargo to the Earth Port would be questionable due to the 36 hours it would take to travel the distance at about 50 knots. However, the range of the US-2 STOL amphibious seaplane is 2,400 nautical miles. It can carry up to 20 passengers and 18 metric tons of cargo. At a cruising speed of 260 knots, the travel time would be about 6 ½ hours, a reasonable time.

As discussed below, a suitable runway (say 5,600 feet long by 150 feet wide)[9] can be built on Howland Island. This would open up the field to a number of U.S. and international commercial aircraft that could become the primary means of moving supplies, cargo and payloads from Honolulu Airport to Howland. With a well-coordinated charter schedule, the use of tub/barge OGVs could be minimized or eliminated. Personnel could travel to/from Howland in corporate jets with speeds up to 550 knots, reducing the travel time to 3 hours.

The remainder of the fleet of watercraft and aircraft described in Section 4.2 would still be required under the Howland Island alternative. However, the facilities for protected moorings, refueling, maintenance and repair would be significantly different than those for the FOP concept described in Section 4.2, as discussed below.

Cargo handling and storage

The FOP concept provides a "mini''-container terminal in the open ocean environment, with all that entails (see Section 4.3.1.) On Howland Island, a fixed cargo berth with sufficient backland for container handling and storage operations could be designed and built that would provide a 2,000 foot wharf for the tug/barge cargo carriers that may arrive from time to time. This would also serve to berth visiting ships including those of the U.S. Coast Guard. As envisioned, this berthing facility would be constructed as a concrete pile-supported wharf parallel to the western shoreline of the island. To save space and avoid duplication, the marine container backland would be integrated with the cargo handling area of the airfield. With the airfield occupying the eastern third of the island, there would sufficient space for a 2,000 foot by 1,500 foot secure open and covered cargo storage and preparation area (69 acres) in the middle of the island. Conventional and specialized container handling equipment described in Section 4.3.1 could be used and maintained within the cargo handling area.

Small craft fleet harbor, service and maintenance facilities

Adjacent to the cargo berths and storage area, along the northwesterly shoreline, a small craft (marine and amphibious aircraft) harbor could be constructed to moor and service the various vessels associated with the operations of the Earth Port and the tether terminus platforms. Instead of the movable, floating wave attenuators associated with the FOP, a fixed breakwater could be

---

[9] These are typical dimensions for general aviation airports that can handle jet engine powered cargo planes. Santa Monica, CA was used as a model.

constructed in an appropriate configuration[10] in the shallow water near the island using pre-fabricated concrete tetrapods resting on an engineered base of local coral and sand. Within the harbor area, a number of floats would provide berthing for the service vessels. Pile-supported fixed berths for high-speed ferries and visiting yachts could also be provided within or adjacent to the harbor as shown in the rendering. Near the berthing area, a series of shop, storage and maintenance buildings would be erected and used in the day-to-day servicing of the various floating vessels. Appropriate refueling stations would be located on the float system. A boat ramp and/or boat lift would be provided from the beach within the protected area of the harbor. Maintenance shops, refueling stations and landing pads for the local service helicopters and aerial drones would be located within the airfield facility.

Meteorological and oceanographic equipment and facilities

This necessary equipment would be similar to that described in Section 4.3.3. Remote sonobuoys and high altitude weather balloons would still be required. Operating these facilities would be simpler from a fixed location rather than aboard a floating platform. More than likely, the weather and oceanographic stations would be located within the work area of the island's airfield.

Engineering facilities

There is sufficient space on the island, probably a central area west of the airfield, to locate and operate a "plant". This will provide electrical power, desalinized water, seawater intakes for fire protection, refrigeration, air conditioning, solid waste collection and disposal system, sewage collection and treatment system, liquid fuel storage and distribution, and other necessary support systems and services for the main operations of the Earth Port facilities on the island and the tether terminus platforms.

Alternative energy considerations

The Howland Island location presents several opportunities for generation of electrical energy that are not as feasible with the FOP concept. Among these opportunities would be nearby wave energy generators, a large solar energy array, wind generators, a small nuclear power plant cooled by ocean water and/or an Ocean Thermal Energy Conversion (OTEC) plant[11]. One or more of these possibilities could be used to augment conventional diesel engine power generation described in the Section 4.3.5.2 of the Study.

The "Village"

All of the hotel facilities and visitor accommodations described in Section 4.3.6 as well as office space, recreational facilities and the local operations and control center could be constructed using conventional techniques and materials in an area adjacent to the small craft harbor on the northwesterly side of the island. This area would serve as a complete "village" for the various personnel assigned to or visiting the Earth Port. It would even have a beach!

---

[10] Design of the breakwater shape and location would be predicated on prevailing wave energy spectrum and offshore water depth contours.

[11] A large-scale test facility for this concept has been built at the Hawaii Natural Energy Lab on the Big Island.

Tether terminus platform operations

Under this island alternative concept, the operations of the tether terminus platforms would remain much the same as those described in the body of the Study. Helicopters would transport personnel to/from the platforms. The offshore service vessels would be moved from their mooring area in the small craft harbor and then loaded with payloads at the container terminal berth(s) on a regular basis. Patrol vessels, aircraft and drones protecting the floating platforms would operate within the secure area around the island.

Comments and conclusions:

The Howland Island alternative concept for the Earth Port seems to present certain advantages over the FOP concept that could reduce development, operation and maintenance costs.. One minor disadvantage is that it is not on the equator. A significant shortcoming is that the platform cannot be moved to avoid storms or space debris. A small shortcoming is the more frequent occurrence of lightning in the area. These shortcomings would be addressed to the System Engineering activity.

The construction and operations described above would have serious detrimental impacts on the habitats of seabirds and other fauna that call Howland Island home. Since the island is managed by USFWS, this issue should be discussed with appropriate U.S. officials at an early stage if it is decided to pursue this concept as a viable alternative for the Earth Port.

www.ingramcontent.com/pod-product-compliance
Lightning Source LLC
Chambersburg PA
CBHW080817170526
45158CB00009B/2455